THE WHALE AND THE REACTOR

The
WHALE
and the
REACTOR

*A Search for Limits in an
Age of High Technology*

LANGDON WINNER

THE UNIVERSITY OF CHICAGO PRESS
Chicago and London

The University of Chicago Press, Chicago 60637
The University of Chicago Press, Ltd., London
© 1986 by The University of Chicago
Paperback edition 1989
All rights reserved. Published 1986
Printed in the United States of America

03 02 01 00 99 98 97   7 8 9 10

Library of Congress Cataloging-in-Publication Data

Winner, Langdon.
    The whale and the reactor.

    Includes index.
    1. Technology—Philosophy.   I. Title.
T14.W54   1986   601   85-8718
ISBN 0-226-90211-0 (paper)

To Gail and Matthew

CONTENTS

PREFACE

THE MAP OF the world shows no country called Technopolis, yet in many ways we are already its citizens. If one observes how thoroughly our lives are shaped by interconnected systems of modern technology, how strongly we feel their influence, respect their authority and participate in their workings, one begins to understand that, like it or not, we have become members of a new order in human history. To an ever-increasing extent, this order of things transcends national boundaries to create roles and relationships grounded in vast, complex instrumentalities of industrial production, electronic communications, transportation, agribusiness, medicine, and warfare. Observing the structures and processes of these vast systems, one begins to comprehend a distinctively modern form of power, the foundations of a technopolitan culture.

The significance of this state of affairs is by no means confined to its material success. When we use terms like "output," "feedback," "interface," and "networking" to express the transactions of everyday life, we reveal how thoroughly artificial things now shape our sense of human being. As we compare our own minds to the operations of a computer, we acknowledge that an understanding of technical devices has somehow merged with the most intimate levels of self-understanding. Seldom, however, are such matters the subject of critical reflection. For most people it is enough to know how technical systems are produced, how they are run, how they are best used, and how they contribute to that vast aggregate of blessings: economic growth.

My aim here is to go further, exploring the meaning of technology for the way we live. What appear to be nothing more than useful instruments are, from another point of view, enduring frameworks of social and political action. How can one look beyond the obvious facts of instrumentality to study the politics of technical objects? Which theoretical perspectives are most helpful in that attempt? In the three chapters of Part I, questions of that kind are examined and some initial steps taken toward developing a political philosophy of technology.

A number of modern social movements have chosen one technology or another as a focus of their hopes or fears. In Part II some of these movements are explored, noting the special opportunities and pitfalls that appear when technology is placed center stage. Appropriate technology, a form of radicalism characteristic of the 1970s, tried to reform society by suggesting we change our tools and our ways of thinking about them. What did the appropriate technologists accomplish? Where did they fall short? For more than a century utopian and anarchist critiques of industrial society have featured political and technical decentralization. While it has wonderful appeal, decentralization turns out to be a very slippery concept. How can it have any importance in a society thoroughly enmeshed in centralized patterns? Many of the passions that have inspired appropriate technology and decentralism have been reborn in the excitement surrounding the so-called computer revolution. Some computer enthusiasts believe that the coming of an information age will inevitably produce a more democratic, egalitarian society and that it will achieve this wonderful condition without the least bit of struggle. I will examine this romantic dream in detail.

A central theme throughout the book concerns the politics of language, a topic that Part III tackles explicitly. Choosing our terms, we express a vision of the world and name our deepest commitments. The quest for political consensus, however, sometimes leads to atrophy of the imagination. In debates about technology, society, and the environment, an extremely narrow range of concepts typically defines the realm of acceptable discussion. For most purposes, issues of efficiency and risk (or some variant of those) are the only ones to receive a thorough hearing. Any broader, deeper, or more perplexing questions are quickly pushed into the shadows and left to wither. How is it that we have gotten stuck packaging some of the important issues that face humanity in such conceptually impoverished terms? What would it take to open up the conversation about technology to include a richer set of cares, categories, and criteria? In the final section we look at three concepts—"nature," "risk," and "values"—to see what light they shed on important choices before us.

In its approach to these matters, this is a work of criticism. If it were literary criticism, everyone would immediately understand that the underlying purpose is positive. A critic of literature examines a work, analyzing its features, evaluating its qualities, seeking a deeper appreciation that might be useful to other readers of the same text. In a similar way, critics of music, theater, and the arts have a valuable, well-established role, serving as a helpful bridge between artists and audiences. Criticism of technology, however, is not yet afforded the same glad welcome. Writers who venture beyond the most pedestrian, dreary conceptions of tools and uses to investigate ways in which technical forms are implicated in the basic patterns and problems of our culture are often greeted with the charge that they are merely "antitechnology" or "blaming technology." All who have recently stepped forward as critics in this realm have been tarred with the same idiot brush, an expression of the desire to stop a much needed dialogue rather than enlarge it. If any readers want to see the present work as "antitechnology," make the most of it. That is their topic, not mine.

What does interest me, however, is identified in the book's subtitle: A Search for Limits. In an age in which the inexhaustible power of scientific technology makes all things possible, it remains to be seen where we will draw the line, where we will be able to say, here are possibilities that wisdom suggests we avoid. I am convinced that any philosophy of technology worth its salt must eventually ask, How can we limit modern technology to match our best sense of who we are and the kind of world we would like to build? In several contexts and variations, that is my question throughout.

All of these are issues in public philosophy, and I have done my best to address them in an open, reasonable, public manner. But they are also extremely personal themes, a fact I do not try to conceal. When the whale surfaces in the final chapter, giving salute to a neighboring reactor, the reader will understand how I came to think about these matters in the first place.

ACKNOWLEDGMENTS

I WISH TO thank the many friends and colleagues whose ideas, suggestions, and criticisms greatly strengthened the research and writing that went into this book. Helpful comments on various parts of the manuscript were offered by Dudley Burton, Peter Euben, John Friedmann, Mary Gibson, Tom Jorling, Rob Kling, Martin Krieger, Frank Laird, Leo Marx, Doug McLean, James O'Connor, John Schaar, Herbert Schiller, Kristin Shrader-Frechette, Merritt Roe Smith, Tim Stroshane, and Sherry Turkle.

I am especially grateful for the companionship of my friends Todd LaPorte, Kai Lee, David Noble, Dick Sclove, Charles Weiner, and Joseph Weizenbaum, whose insights into the human dimensions of technology have been a continuing inspiration over the years. The generous encouragement of Stan Draenos, Richard Farson, Andrew Feenberg, Richard Gordon, Greil Marcus, James Miller, Bruce Miroff, and Don Van Vliet has often boosted my spirits when I needed it most.

My special thanks also to Mary Kiegelis and Astrid Kuhr for their help in preparing the final manuscript.

The early stages of my research were financed by a fellowship from the National Endowment for the Humanities. The book was completed with funds from the Sustained Development Award of the Program on Ethics and Values of the National Science Foundation and the National Endowment for the Humanities, OSS-8018089. Its views are those of the author and do not necessarily reflect those of the NSF or NEH. I wish to express my warm thanks to these agencies for their strong support.

Most of all, I owe a debt of gratitude to my wife, Gail Stuart, whose lively spirit brings joy to my life and work.

Earlier versions of several chapters appeared in other publications, as follows: chapter 1, as "Technologies as Forms of Life," in *Epistemology, Methodology, and the Social Sciences*, R. S. Cohen and M. W. Wartofsky (eds.) (Dordrecht: D. Reidel, 1983), 249–263, © 1983 by D. Reidel Publishing Company; chapter 2, as "Do Artifacts Have Politics?" in *Daedalus* 109, no. 1 (Winter 1980): 121–136; chapter 3, as "Technē and Politeia: The Technical Constitution of Society," in *Philosophy and Technology*, Paul T. Durbin and Fredrich Rapp (eds.) (Dordrecht: D. Reidel, 1983), 97–111, © 1983 by D. Reidel Publishing Company; chapter 4, as "Building the Better Mousetrap: Appropriate Technology as a Social Movement," in *Appropriate Technology and Social Values: A Critical Appraisal*, Franklin A. Long and Alexandra Oleson (eds.) (Cambridge: Ballinger, 1980), 27–51, © 1980 by The American Academy of Arts and Sciences; chapter 5, as "Decentralization: Its Meaning in Politics and Material Culture," in *Research in Philosophy and Technology*, Paul T. Durbin (ed.) (Greenwich, Conn.: JAI Press, 1983), 43–52; chapter 6, as "Mythinformation: Romantic Politics in the Computer Revolution," in *Research in Philosophy and Technology*, Paul T. Durbin (ed.) (Greenwich, Conn.: JAI Press, 1984), 287–304; chapter 8, as "On Not Hitting the Tar-Baby: Risk Assessment and Conservatism," in *To Breathe Freely: Risk, Consent, and Air*, Mary Gibson (ed.) (Totowa, N.J.: Rowman and Allanheld, 1985); and chapter 10, as "The Whale and the Reactor: A Personal Memoir," in *The Journal of American Culture* 3, no. 3 (Fall 1980): 446–455.

I

A PHILOSOPHY OF TECHNOLOGY

1

TECHNOLOGIES AS
FORMS OF LIFE

FROM THE EARLY DAYS of manned space travel comes a story
that exemplifies what is most fascinating about the human encounter with modern technology. Orbiting the earth aboard
Friendship 7 in February 1962, astronaut John Glenn noticed
something odd. His view of the planet was virtually unique in
human experience; only Soviet pilots Yuri Gagarin and Gherman
Titov had preceded him in orbital flight. Yet as he watched the
continents and oceans moving beneath him, Glenn began to feel
that he had seen it all before. Months of simulated space shots in
sophisticated training machines and centifuges had affected his
ability to respond. In the words of chronicler Tom Wolfe, "The
world demanded awe, because this was a voyage through the
stars. But he couldn't feel it. The backdrop of the event, the
stage, the environment, the true orbit . . . was not the vast
reaches of the universe. It was the simulators. *Who could possibly
understand this?*"[1] Synthetic conditions generated in the training
center had begun to seem more "real" than the actual experience.

It is reasonable to suppose that a society thoroughly committed to making artificial realities would have given a great deal
of thought to the nature of that commitment. One might expect, for example, that the philosophy of technology would be a
topic widely discussed by scholars and technical professionals, a
lively field of inquiry often chosen by students at our universities and technical institutes. One might even think that the
basic issues in this field would be well defined, its central controversies well worn. However, such is not the case. At this late
date in the development of our industrial/technological civiliza-

3

tion the most accurate observation to be made about the philosophy of technology is that there really isn't one.

The basic task for a philosophy of technology is to examine critically the nature and significance of artificial aids to human activity. That is its appropriate domain of inquiry, one that sets it apart from, say, the philosophy of science. Yet if one turns to the writings of twentieth-century philosophers, one finds astonishingly little attention given to questions of that kind. The six-volume *Encyclopedia of Philosophy*, a recent compendium of major themes in various traditions of philosophical discourse, contains no entry under the category "technology."[2] Neither does that work contain enough material under possible alternative headings to enable anyone to piece together an idea of what a philosophy of technology might be.

True, there are some writers who have taken up the topic. The standard bibliography in the philosophy of technology lists well over a thousand books and articles in several languages by nineteenth- and twentieth-century authors.[3] But reading through the material listed shows, in my view, little of enduring substance. The best writing on this theme comes to us from a few powerful thinkers who have encountered the subject in the midst of much broader and ambitious investigations—for example, Karl Marx in the development of his theory of historical materialism or Martin Heidegger as an aspect of his theory of ontology. It may be, in fact, that the philosophy is best seen as a derivative of more fundamental questions. For despite the fact that nobody would deny its importance to an adequate understanding of the human condition, technology has never joined epistemology, metaphysics, esthetics, law, science, and politics as a fully respectable topic for philosophical inquiry.

Engineers have shown little interest in filling this void. Except for airy pronouncements in yearly presidential addresses at various engineering societies, typically ones that celebrate the contributions of a particular technical vocation to the betterment of humankind, engineers appear unaware of any philosophical questions their work might entail. As a way of starting a conversation with my friends in engineering, I sometimes ask, "What are the founding principles of your discipline?" The question is always greeted with puzzlement. Even when I explain what I am after, namely, a coherent account of the nature and significance of the branch of engineering in which they are involved, the question still means nothing to them. The scant few who raise important first questions about their technical professions are usually seen by their colleagues as dangerous cranks and radicals. If Socrates' suggestion that the "unexamined life is not worth living" still holds, it is news to most engineers.[4]

Technological Somnambulism

WHY IS IT that the philosophy of technology has never really gotten under way? Why has a culture so firmly based upon countless sophisticated instruments, techniques, and systems remained so steadfast in its reluctance to examine its own foundations? Much of the answer can be found in the astonishing hold the idea of "progress" has exercised on social thought during the industrial age. In the twentieth century it is usually taken for granted that the only reliable sources for improving the human condition stem from new machines, techniques, and chemicals. Even the recurring environmental and social ills that have accompanied technological advancement have rarely dented this faith. It is still a prerequisite that the person running for public office swear his or her unflinching confidence in a positive link between technical development and human well-being and affirm that the next wave of innovations will surely be our salvation.

There is, however, another reason why the philosophy of technology has never gathered much steam. According to conventional views, the human relationship to technical things is too obvious to merit serious reflection. The deceptively reasonable notion that we have inherited from much earlier and less complicated times divides the range of possible concerns about technology into two basic categories: *making* and *use*. In the first of these our attention is drawn to the matter of "how things work" and of "making things work." We tend to think that this is a fascination of certain people in certain occupations, but not for anyone else. "How things work" is the domain of inventors, technicians, engineers, repairmen, and the like who prepare artificial aids to human activity and keep them in good working order. Those not directly involved in the various spheres of "making" are thought to have little interest in or need to know about the materials, principles, or procedures found in those spheres.

What the others do care about, however, are tools and uses.

This is understood to be a straightforward matter. Once things have been made, we interact with them on occasion to achieve specific purposes. One picks up a tool, uses it, and puts it down. One picks up a telephone, talks on it, and then does not use it for a time. A person gets on an airplane, flies from point A to point B, and then gets off. The proper interpretation of the meaning of technology in the mode of use seems to be nothing more complicated than an occasional, limited, and nonproblematic interaction.

The language of the notion of "use" also includes standard terms that enable us to interpret technologies in a range of moral contexts. Tools can be "used well or poorly" and for "good or bad purposes"; I can use my knife to slice a loaf of bread or to stab the next person that walks by. Because technological objects and processes have a promiscuous utility, they are taken to be fundamentally neutral as regards their moral standing.

The conventional idea of what technology is and what it means, an idea powerfully reinforced by familiar terms used in everyday language, needs to be overcome if a critical philosophy of technology is to move ahead. The crucial weakness of the conventional idea is that it disregards the many ways in which technologies provide structure for human activity. Since, according to accepted wisdom, patterns that take shape in the sphere of "making" are of interest to practitioners alone, and since the very essence of "use" is its occasional, innocuous, nonstructuring occurrence, any further questioning seems irrelevant.[5]

If the experience of modern society shows us anything, however, it is that technologies are not merely aids to human activity, but also powerful forces acting to reshape that activity and its meaning. The introduction of a robot to an industrial workplace not only increases productivity, but often radically changes the process of production, redefining what "work" means in that setting. When a sophisticated new technique or instrument is adopted in medical practice, it transforms not only what doctors do, but also the ways people think about health, sickness, and medical care. Widespread alterations of this kind in techniques of communication, transportation, manufacturing, agriculture, and the like are largely what distinguishes our times from early periods of human history. The kinds of things we are apt to see as "mere" technological entities become much more interesting and problematic if we begin to observe how broadly they are involved in conditions of social and moral life.

6

It is true that recurring patterns of life's activity (whatever their origins) tend to become unconscious processes taken for granted. Thus, we do not pause to reflect upon how we speak a language as we are doing so or the motions we go through in taking a shower. There is, however, one point at which we may become aware of a pattern taking shape—the very first time we encounter it. An opportunity of that sort occurred several years ago at the conclusion of a class I was teaching. A student came to my office on the day term papers were due and told me his essay would be late. "It crashed this morning," he explained. I immediately interpreted this as a "crash" of the conceptual variety, a flimsy array of arguments and observations that eventually collapses under the weight of its own ponderous absurdity. Indeed, some of my own papers have "crashed" in exactly that manner. But this was not the kind of mishap that had befallen this particular fellow. He went on to explain that his paper had been composed on a computer terminal and that it had been stored in a time-sharing minicomputer. It sometimes happens that the machine "goes down" or "crashes," making everything that happens in and around it stop until the computer can be "brought up," that is, restored to full functioning.

As I listened to the student's explanation, I realized that he was telling me about the facts of a particular form of activity in modern life in which he and others similarly situated were already involved and that I had better get ready for. I remembered J. L. Austin's little essay "A Plea for Excuses" and noticed that the student and I were negotiating one of the boundaries of contemporary moral life—where and how one gives and accepts an excuse in a particular technology-mediated situation.[6] He was, in effect, asking me to recognize a new world of parts and pieces and to acknowledge appropriate practices and expectations that hold in that world. From then on, a knowledge of this situation would be included in my understanding of not only "how things work" in that generation of computers, but also how we do things as a consequence, including which rules to follow when the machines break down. Shortly thereafter I got used to computers crashing, disrupting hotel reservations, banking, and other everyday transactions; eventually, my own papers began crashing in this new way.

Some of the moral negotiations that accompany technological change eventually become matters of law. In recent times, for example, a number of activities that employ computers as their operating medium have been legally defined as "crimes." Is unauthorized access to a computerized data base a criminal offense? Given the fact that electronic information is in the strictest sense intangible, under what conditions is it "property" subject to theft? The law has had to stretch and reorient its traditional categories to encompass such problems, creating whole new classes of offenses and offenders.

The ways in which technical devices tend to engender distinctive worlds of their own can be seen in a more familiar case. Picture two men traveling in the same direction along a street on a peaceful, sunny day, one of them afoot and the other driving an automobile. The pedestrian has a certain flexibility of movement: he can pause to look in a shop window, speak to passersby, and reach out to pick a flower from a sidewalk garden. The driver, although he has the potential to move much faster, is constrained by the enclosed space of the automobile, the physical dimensions of the highway, and the rules of the road. His realm is spatially structured by his intended destination, by a periphery of more-or-less irrelevant objects (scenes for occasional side glances), and by more important objects of various kinds—moving and parked cars, bicycles, pedestrians, street signs, etc., that stand in his way. Since the first rule of good driving is to avoid hitting things, the immediate environment of the motorist becomes a field of obstacles.

Imagine a situation in which the two persons are next-door neighbors. The man in the automobile observes his friend strolling along the street and wishes to say hello. He slows down, honks his horn, rolls down the window, sticks out his head, and shouts across the street. More likely than not the pedestrian will be startled or annoyed by the sound of the horn. He looks around to see what's the matter and tries to recognize who can be yelling at him across the way. "Can you come to dinner Saturday night?" the driver calls out over the street noise. "What?" the pedestrian replies, straining to understand. At that moment another car to the rear begins honking to break up the temporary traffic jam. Unable to say anything more, the driver moves on.

What we see here is an automobile collision of sorts, although not one that causes bodily injury. It is a collision between the *world* of the driver and that of the pedestrian. The attempt to extend a greeting and invitation, ordinarily a simple gesture, is complicated by the presence of a technological device and its standard operating conditions. The communication between the two men is shaped by an incompatibility of the form of locomotion known as walking and a much newer one, automobile driving. In cities such as Los Angeles, where the physical landscape and prevailing social habits assume everyone drives a car, the simple act of walking can be cause for alarm. The U.S. Supreme Court decided one case involving a young man who enjoyed taking long walks late at night through the streets of San Diego and was repeatedly arrested by police as a suspicious character. The Court decided in favor of the pedestrian, noting that he had not been engaged in burglary or any other illegal act. Merely traveling by foot is not yet a crime.[7]

Knowing how automobiles are made, how they operate, and how they are used and knowing about traffic laws and urban transportation policies does little to help us understand how automobiles affect the texture of modern life. In such cases a strictly instrumental/functional understanding fails us badly. What is needed is an interpretation of the ways, both obvious and subtle, in which everyday life is transformed by the mediating role of technical devices. In hindsight the situation is clear to everyone. Individual habits, perceptions, concepts of self, ideas of space and time, social relationships, and moral and political boundaries have all been powerfully restructured in the course of modern technological development. What is fascinating about this process is that societies involved in it have quickly altered some of the fundamental terms of human life without appearing to do so. Vast transformations in the structure of our common world have been undertaken with little attention to what those alterations mean. Judgments about technology have been made on narrow grounds, paying attention to such matters as whether a new device serves a particular need, performs more efficiently than its predecessor, makes a profit, or provides a convenient service. Only later does the broader significance of the choice become clear, typically as a series of surprising "side effects" or "secondary consequences." But it seems characteristic of our culture's involvement with technology that we are seldom inclined to examine, discuss, or judge pending innovations with broad, keen awareness of what those changes mean. In the technical realm we repeatedly enter into a series of social contracts, the terms of which are revealed only after the signing.

It may seem that the view I am suggesting is that of technological determinism: the idea that technological innovation is the basic cause of changes in society and that human beings have little choice other than to sit back and watch this ineluctable process unfold. But the concept of determinism is much too strong, far too sweeping in its implications to provide an adequate theory. It does little justice to the genuine choices that arise, in both principle and practice, in the course of technical and social transformation. Being saddled with it is like attempting to describe all instances of sexual intercourse based only on the concept of rape. A more revealing notion, in my view, is that of technological somnambulism. For the interesting puzzle in our times is that we so willingly sleepwalk through the process of reconstituting the conditions of human existence.

Beyond Impacts and Side Effects

SOCIAL SCIENTISTS have tried to awaken the sleeper by developing methods of technology assessment. The strength of these methods is that they shed light on phenomena that were previously overlooked. But an unfortunate shortcoming of technology assessment is that it tends to see technological change as a "cause" and everything that follows as an "effect" or "impact." The role of the researcher is to identify, observe, and explain these effects. This approach assumes that the causes have already occurred or are bound to do so in the normal course of events. Social research boldly enters the scene to study the "consequences" of the change. After the bulldozer has rolled over us, we can pick ourselves up and carefully measure the treadmarks. Such is the impotent mission of technological "impact" assessment.

A somewhat more farsighted version of technology assessment is sometimes used to predict which changes are likely to happen, the "social impacts of computerization" for example. With these forecasts at its disposal, society is, presumably, better able to chart its course. But, once again, the attitude in which the predictions are offered usually suggests that the "impacts" are going to happen in any case. Assertions of the sort "Computerization will bring about a revolution in the way we educate our children" carry the strong implication that those who will experience the change are obliged simply to endure it. Humans must adapt. That is their destiny. There is no tampering with the source of change, and only minor modifications are possible

10

at the point of impact (perhaps some slight changes in the fashion contour of this year's treadmarks).

But we have already begun to notice another view of technological development, one that transcends the empirical and moral shortcomings of cause-and-effect models. It begins with the recognition that as technologies are being built and put to use, significant alterations in patterns of human activity and human institutions are already taking place. New worlds are being made. There is nothing "secondary" about this phenomenon. It is, in fact, the most important accomplishment of any new technology. The construction of a technical system that involves human beings as operating parts brings a reconstruction of social roles and relationships. Often this is a result of a new system's own operating requirements: it simply will not work unless human behavior changes to suit its form and process. Hence, the very act of using the kinds of machines, techniques, and systems available to us generates patterns of activities and expectations that soon become "second nature." We do indeed "use" telephones, automobiles, electric lights, and computers in the conventional sense of picking them up and putting them down. But our world soon becomes one in which telephony, automobility, electric lighting, and computing are forms of life in the most powerful sense: life would scarcely be thinkable without them.

My choice of the term "forms of life" in this context derives from Ludwig Wittgenstein's elaboration of that concept in *Philosophical Investigations*. In his later writing Wittgenstein sought to overcome an extremely narrow view of the structure of language then popular among philosophers, a view that held language to be primarily a matter of naming things and events. Pointing to the richness and multiplicity of the kinds of expression or "language games" that are a part of everyday speech, Wittgenstein argued that "the speaking of language is a part of an activity, or of a form of life."[8] He gave a variety of examples—the giving of orders, speculating about events, guessing riddles, making up stories, forming and testing hypotheses, and so forth—to indicate the wide range of language games involved in various "forms of life." Whether he meant to suggest that these are patterns that occur naturally to all human beings or that they are primarily cultural conventions that can change with time and setting is a question open to dispute.[9] For the purposes here, what matters is not the ultimate philosophical status

11

of Wittgenstein's concept but its suggestiveness in helping us to overcome another widespread and extremely narrow conception: our normal understanding of the meaning of technology in human life.

As they become woven into the texture of everyday existence, the devices, techniques, and systems we adopt shed their tool-like qualities to become part of our very humanity. In an important sense we become the beings who work on assembly lines, who talk on telephones, who do our figuring on pocket calculators, who eat processed foods, who clean our homes with powerful chemicals. Of course, working, talking, figuring, eating, cleaning, and such things have been parts of human activity for a very long time. But technological innovations can radically alter these common patterns and on occasion generate entirely new ones, often with surprising results. The role television plays in our society offers some poignant examples. None of those who worked to perfect the technology of television in its early years and few of those who brought television sets into their homes ever intended the device to be employed as the universal babysitter. That, however, has become one of televisions' most common functions in the modern home. Similarly, if anyone in the 1930s had predicted people would eventually be watching seven hours of television each day, the forecast would have been laughed away as absurd. But recent surveys indicate that we Americans do spend that much time, roughly one-third of our lives, staring at the tube. Those who wish to reassert freedom of choice in the matter sometimes observe, "You can always turn off your TV." In a trivial sense that is true. At least for the time being the on/off button is still included as standard equipment on most sets (perhaps someday it will become optional). But given how central television has become to the content of everyday life, how it has become the accustomed topic of conversation in workplaces, schools, and other social gatherings, it is apparent that television is a phenomenon that, in the larger sense, cannot be "turned off" at all. Deeply insinuated into people's perceptions, thoughts, and behavior, it has become an indelible part of modern culture.

Most changes in the content of everyday life brought on by technology can be recognized as versions of earlier patterns. Parents have always had to entertain and instruct children and to find ways of keeping the little ones out of their hair. Having youngsters watch several hours of television cartoons is, in one way of looking at the matter, merely a new method for handling this age-old task, although the "merely" is of no small significance. It is important to ask, Where, if at all, have modern technologies added *fundamentally new* activities to the range of things human beings do? Where and how have innovations in science and technology begun to alter the very *conditions of life* itself? Is computer programming only a powerful recombination of forms of life known for ages—doing mathematics, listing, sorting, planning, organizing, etc.—or is it something unprecedented? Is industrialized agribusiness simply a renovation of older ways of farming, or does it amount to an entirely new phenomenon?

Certainly, there are some accomplishments of modern technology, manned air flight, for example, that are clearly altogether novel. Flying in airplanes is not just another version of modes of travel previously known; it is something new. Although the hope of humans flying is as old as the myth of Daedalus and Icarus or the angels of the *Old Testament*, it took a certain kind of modern machinery to realize the dream in practice. Even beyond the numerous breakthroughs that have pushed the boundaries of human action, however, lie certain kinds of changes now on the horizon that would amount to a fundamental change in the conditions of human life itself. One such prospect is that of altering human biology through genetic engineering. Another is the founding of permanent settlements in outer space. Both of these possibilities call into question what it means to be human and what constitutes "the human condition."[10] Speculation about such matters is now largely the work of science fiction, whose notorious perversity as a literary genre signals the troubles that lie in wait when we begin thinking about becoming creatures fundamentally different from any the earth has seen. A great many futuristic novels are blatantly technopornographic.

But, on the whole, most of the transformations that occur in the wake of technological innovation are actually variations of very old patterns. Wittgenstein's philosophically conservative maxim "What has to be accepted, the given, is—so one could say—*forms of life*" could well be the guiding rule of a phenomenology of technical practice.[11] For instance, asking a question and awaiting an answer, a form of interaction we all know well, is much the same activity whether it is a person we are confronting or a computer. There are, of course, significant differences between persons and computers (although it is fashionable in some circles to ignore them). Forms of life that we mastered before the coming of the computer shape our expectations as we begin to use the instrument. One strategy of software design, therefore, tries to "humanize" the computers by having them say "Hello" when the user logs in or having them respond with witty remarks when a person makes an error. We carry with us highly structured anticipations about entities that appear to participate, if only minimally, in forms of life and associated language games that are parts of human culture. Those anticipations provide much of the persuasive power of those who prematurely claim great advances in "artificial intelligence" based on narrow but impressive demonstrations of computer performance. But then children have always fantasized that their dolls were alive and talking.

The view of technologies as forms of life I am proposing has its clearest beginnings in the writings of Karl Marx. In Part I of *The German Ideology*, Marx and Engels explain the relationship of human individuality and material conditions of production as follows: "The way in which men produce their means of subsistence depends first of all on the nature of the means of subsistence they actually find in existence and have to reproduce. This mode of production must not be considered simply as being the reproduction of the physical existence of the individuals. Rather it is a definite form of activity of these individuals, a definite form of expressing their life, a definite *mode of life* on their part. As individuals express their life, so they are."[12]

Marx's concept of production here is a very broad and suggestive one. It reveals the total inadequacy of any interpretation that finds social change a mere "side effect" or "impact" of technological innovation. While he clearly points to means of production that sustain life in an immediate, physical sense, Marx's view extends to a general understanding of human development in a world of diverse natural resources, tools, machines, products, and social relations. The notion is clearly not one of occasional human interaction with devices and material conditions that leave individuals unaffected. By changing the shape of material things, Marx observes, we also change ourselves. In this process human beings do not stand at the mercy of a great deterministic punch press that cranks out precisely tailored persons at a certain rate during a given historical period. Instead, the situation Marx describes is one in which individuals are actively involved in the daily creation and recreation, production and reproduction of the world in which they live. Thus, as they employ tools and techniques, work in social labor arrangements, make and consume products, and adapt their behavior to the material conditions they encounter in their natural and artificial environment, individuals realize possibilities for human existence that are inaccessible in more primitive modes of production.

Marx expands upon this idea in "The Chapter on Capital" in the *Grundrisse*. The development of forces of production in history, he argues, holds the promise of the development of a many-sided individuality in all human beings. Capital's unlimited pursuit of wealth leads it to develop the productive powers of labor to a state "where the possession and preservation of general wealth require a lesser labour time of society as a whole, and where the labouring society relates scientifically to the process of its progressive reproduction, its reproduction in constantly greater abundance." This movement toward a general form of wealth "creates the material elements for the development of the rich individuality which is all-sided in its production as in its consumption, and whose labour also therefore appears no longer as labour, but as the full development of activity itself."[13]

If one has access to tools and materials of woodworking, a person can develop the human qualities found in the activities of carpentry. If one is able to employ the instruments and techniques of music making, one can become (in that aspect of one's life) a musician. Marx's ideal here, a variety of materialist humanism, anticipates that in a properly structured society under modern conditions of production, people would engage in a very wide range of activities that enrich their individuality along many dimensions. It is that promise which, he argues, the institutions of capitalism thwart and cripple.[14]

As applied to an understanding of technology, the philosophies of Marx and Wittgenstein direct our attention to the fabric of everyday existence. Wittgenstein points to a vast multiplicity of cultural practices that comprise our common world. Asking us to notice "what we say when," his approach can help us recognize the way language reflects the content of technical practice. It makes sense to ask, for example, how the adoption of digital computers might alter the way people think of their own faculties and activities. If Wittgenstein is correct, we would expect that changes of this kind would appear, sooner or later, in the language people use to talk about themselves. Indeed, it has now become commonplace to hear people say "I need to access your data." "I'm not programmed for that." "We must improve our interface." "The mind is the best computer we have."

Marx, on the other hand, recommends that we see the actions and interactions of everyday life within an enormous tapestry of historical developments. On occasion, as in the chapter on "Machinery and Large-Scale Industry" in *Capital*, his mode of interpretation also includes a place for a more microscopic treatment of specific technologies in human experience.[15] But on the whole his theory seeks to explain very large patterns, especially relationships between different social classes, that unfold at each stage in the history of material production. These developments set the stage for people's ability to survive and express themselves, for their ways of being human.

Return to Making

TO INVOKE Wittgenstein and Marx in this context, however, is not to suggest that either one or both provide a sufficient basis for a critical philosophy of technology. Proposing an attitude in which forms of life must be accepted as "the given," Wittgenstein decides that philosophy "leaves everything as it is."[16] Although some Wittgensteinians are eager to point out that this position does not necessarily commit the philosopher to conservatism in an economic or political sense, it does seem that as applied to the study of forms of life in the realm of technology, Wittgenstein leaves us with little more than a passive traditionalism. If one hopes to interpret technological phenomena in a way that suggests positive judgments and actions, Wittgensteinian philosophy leaves much to be desired.

In a much different way Marx and Marxism contain the potential for an equally woeful passivity. This mode of understanding places its hope in historical tendencies that promise human emancipation at some point. As forces of production and social relations of production develop and as the proletaria' makes its way toward revolution, Marx and his orthodox followers are willing to allow capitalist technology, for example, the factory system, to develop to its farthest extent. Marx and Engels scoffed at the utopians, anarchists, and romantic critics of industrialism who thought it possible to make moral and political judgments about the course a technological society ought to take and to influence that path through the application of philosophical principles. Following this lead, most Marxists have believed that while capitalism is a target to be attacked, technological expansion is entirely good in itself, something to be encouraged without reservation. In its own way, then, Marxist theory upholds an attitude as nearly lethargic as the Wittgensteinian decision to "leave everything as it is." The famous eleventh thesis on Feuerbach—"The philosophers have only interpreted the world in various ways; the point, however, is to change it"—conceals an important qualification: that judgment, action, and change are ultimately products of history. In its view of technological development Marxism anticipates a history of rapidly evolving material productivity, an inevitable course of events in which attempts to propose moral and political limits have no place. When socialism replaces capitalism, so the promise goes, the machine will finally move into high gear, presumably releasing humankind from its age-old miseries.

Whatever their shortcomings, however, the philosophies of Marx and Wittgenstein share a fruitful insight: the observation that social activity is an ongoing process of world-making. Throughout their lives people come together to renew the fabric of relationships, transactions, and meanings that sustain their common existence. Indeed, if they did not engage in this continuing activity of material and social production, the human world would literally fall apart. All social roles and frameworks—from the most rewarding to the most oppressive— must somehow be restored and reproduced with the rise of the sun each day.

From this point of view, the important question about technology becomes, As we "make things work," what kind of *world* are we making? This suggests that we pay attention not only to the making of physical instruments and processes, although that certainly remains important, but also to the production of psychological, social, and political conditions as a part of any significant technical change. Are we going to design and build circumstances that enlarge possibilities for growth in human freedom, sociability, intelligence, creativity, and self-government? Or are we headed in an altogether different direction?

It is true that not every technological innovation embodies choices of great significance. Some developments are more-or-less innocuous; many create only trivial modifications in how we live. But in general, where there are substantial changes

17

being made in what people are doing and at a substantial investment of social resources, then it always pays to ask in advance about the qualities of the artifacts, institutions, and human experiences currently on the drawing board.

Inquiries of this kind present an important challenge to all disciplines in the social sciences and humanities. Indeed, there are many historians, anthropologists, sociologists, psychologists, and artists whose work sheds light on long-overlooked human dimensions of technology. Even engineers and other technical professionals have much to contribute here when they find courage to go beyond the narrow-gauge categories of their training.

The study of politics offers its own characteristic route into this territory. As the political imagination confronts technologies as forms of life, it should be able to say something about the choices (implicit or explicit) made in the course of technological innovation and the grounds for making those choices wisely. That is a task I take up in the next two chapters. Through technological creation and many other ways as well, we make a world for each other to live in. Much more than we have acknowledged in the past, we must admit our responsibility for what we are making.

2

(Human made items)

DO ARTIFACTS HAVE
POLITICS?

NO IDEA IS more provocative in controversies about technology and society than the notion that technical things have political qualities. At issue is the claim that the machines, structures,
and systems of modern material culture can be accurately judged
not only for their contributions to efficiency and productivity *how we think about them*
and their positive and negative environmental side effects, but
also for the ways in which they can embody specific forms of *politics*
power and authority. Since ideas of this kind are a persistent and
troubling presence in discussions about the meaning of technology, they deserve explicit attention.

Writing in the early 1960s, Lewis Mumford gave classic statement to one version of the theme, arguing that "from late neolithic times in the Near East, right down to our own day, two
technologies have recurrently existed side by side: one authoritarian, the other democratic, the first system-centered, im-
mensely powerful, but inherently unstable, the other man-
centered, relatively weak, but resourceful and durable."[1] This
thesis stands at the heart of Mumford's studies of the city, architecture, and history of technics, and mirrors concerns voiced earlier in the works of Peter Kropotkin, William Morris, and other
nineteenth-century critics of industrialism. During the 1970s,
antinuclear and pro-solar energy movements in Europe and the
United States adopted a similar notion as the centerpiece of their
arguments. According to environmentalist Denis Hayes, "The
increased deployment of nuclear power facilities must lead society toward authoritarianism. Indeed, safe reliance upon nuclear power as the principal source of energy may be possible
only in a totalitarian state." Echoing the views of many propo-

19

nents of appropriate technology and the soft energy path, Hayes contends that "dispersed solar sources are more compatible than centralized technologies with social equity, freedom and cultural pluralism."[2]

An eagerness to interpret technical artifacts in political language is by no means the exclusive property of critics of large-scale, high-technology systems. A long lineage of boosters has insisted that the biggest and best that science and industry made available were the best guarantees of democracy, freedom, and social justice. The factory system, automobile, telephone, radio, television, space program, and of course nuclear power have all at one time or another been described as democratizing, liberating forces. David Lillienthal's *T.V.A.: Democracy on the March*, for example, found this promise in the phosphate fertilizers and electricity that technical progress was bringing to rural Americans during the 1940s.[3] Three decades later Daniel Boorstin's *The Republic of Technology* extolled television for "its power to disband armies, to cashier presidents, to create a whole new democratic world—democratic in ways never before imagined, even in America."[4] Scarcely a new invention comes along that someone doesn't proclaim it as the salvation of a free society.

It is no surprise to learn that technical systems of various kinds are deeply interwoven in the conditions of modern politics. The physical arrangements of industrial production, warfare, communications, and the like have fundamentally changed the exercise of power and the experience of citizenship. But to go beyond this obvious fact and to argue that certain technologies *in themselves* have political properties seems, at first glance, completely mistaken. We all know that people have politics; things do not. To discover either virtues or evils in aggregates of steel, plastic, transistors, integrated circuits, chemicals, and the like seems just plain wrong, a way of mystifying human artifice and of avoiding the true sources, the human sources of freedom and oppression, justice and injustice. Blaming the hardware appears even more foolish than blaming the victims when it comes to judging conditions of public life.

Hence, the stern advice commonly given those who flirt with the notion that technical artifacts have political qualities: What matters is not technology itself, but the social or economic system in which it is embedded. This maxim, which in a number of variations is the central premise of a theory that can be called the social determination of technology, has an obvious wisdom. It serves as a needed corrective to those who focus uncritically upon such things as "the computer and its social impacts" but who fail to look behind technical devices to see the social circumstances of their development, deployment, and use. This view provides an antidote to naive technological determinism— the idea that technology develops as the sole result of an internal dynamic and then, unmediated by any other influence, molds society to fit its patterns. Those who have not recognized the ways in which technologies are shaped by social and economic forces have not gotten very far.

But the corrective has its own shortcomings; taken literally, it suggests that technical *things* do not matter at all. Once one has done the detective work necessary to reveal the social origins— power holders behind a particular instance of technological change—one will have explained everything of importance. This conclusion offers comfort to social scientists. It validates what they had always suspected namely, that there is nothing distinctive about the study of technology in the first place. Hence, they can return to the standard models of social power— those of interest-group politics, bureaucratic politics, Marxist models of class struggle, and the like—and have everything they need. The social determination of technology is, in this view, essentially no different from the social determination of, say, welfare policy or taxation.

There are, however, good reasons to believe that technology is politically significant in its own right, good reasons why the standard models of social science only go so far in accounting for what is most interesting and troublesome about the subject. Much of modern social and political thought contains recurring statements of what can be called a theory of technological politics, an odd mongrel of notions often crossbred with orthodox liberal, conservative, and socialist philosophies.[5] The theory of technological politics draws attention to the momentum of large-scale sociotechnical systems, to the response of modern societies to certain technological imperatives, and to the ways human ends are powerfully transformed as they are adapted to technical means. This perspective offers a novel framework of interpretation and explanation for some of the more puzzling patterns that have taken shape in and around the growth of modern material culture. Its starting point is a decision to take technical artifacts seriously. Rather than insist that we immediately reduce everything to the interplay of social forces, the theory of technological politics suggests that we pay attention to the characteristics of technical objects and the meaning of those characteristics. A necessary complement to, rather than a replacement for, theories of the social determination of technology, this approach identifies certain technologies as political phenomena in their own right. It points us back, to borrow Edmund Husserl's philosophical injunction, *to the things themselves.*

In what follows I will outline and illustrate two ways in which artifacts can contain political properties. First are instances in which the invention, design, or arrangement of a specific technical device or system becomes a way of settling an issue in the affairs of a particular community. Seen in the proper light, examples of this kind are fairly straightforward and easily understood. Second are cases of what can be called "inherently political technologies," man-made systems that appear to require or to be strongly compatible with particular kinds of political relationships. Arguments about cases of this kind are much more troublesome and closer to the heart of the matter. By the term "politics" I mean arrangements of power and authority in human associations as well as the activities that take place within those arrangements. For my purposes here, the term "technology" is understood to mean all of modern practical artifice, but to avoid confusion I prefer to speak of "technologies" plural, smaller or larger pieces or systems of hardware of a specific kind.[6] My intention is not to settle any of the issues here once and for all, but to indicate their general dimensions and significance.

Technical Arrangements and Social Order

ANYONE WHO has traveled the highways of America and has gotten used to the normal height of overpasses may well find something a little odd about some of the bridges over the parkways on Long Island, New York. Many of the overpasses are extraordinarily low, having as little as nine feet of clearance at the curb. Even those who happened to notice this structural peculiarity would not be inclined to attach any special meaning to it. In our accustomed way of looking at things such as roads and bridges, we see the details of form as innocuous and seldom give them a second thought.

It turns out, however, that some two hundred or so low-hanging overpasses on Long Island are there for a reason. They were deliberately designed and built that way by someone who wanted to achieve a particular social effect. Robert Moses, the master builder of roads, parks, bridges, and other public works of the 1920s to the 1970s in New York, built his overpasses according to specifications that would discourage the presence of buses on his parkways. According to evidence provided by Moses' biographer, Robert A. Caro, the reasons reflect Moses' social class bias and racial prejudice. Automobile-owning whites of "upper" and "comfortable middle" classes, as he called them, would be free to use the parkways for recreation and commuting. Poor people and blacks, who normally used public transit, were kept off the roads because the twelve-foot tall buses could not handle the overpasses. One consequence was to limit access of racial minorities and low-income groups to Jones Beach, Moses' widely acclaimed public park. Moses made doubly sure of this result by vetoing a proposed extension of the Long Island Railroad to Jones Beach.

Robert Moses' life is a fascinating story in recent U.S. political history. His dealings with mayors, governors, and presidents; his careful manipulation of legislatures, banks, labor unions, the press, and public opinion could be studied by political scientists for years. But the most important and enduring results of his work are his technologies, the vast engineering projects that give New York much of its present form. For generations after Moses' death and the alliances he forged have fallen apart, his public works, especially the highways and bridges he built to favor the use of the automobile over the development of mass transit, will continue to shape that city. Many of his monumental structures of concrete and steel embody a systematic social inequality, a way of engineering relationships among people that, after a time, became just another part of the landscape. As New York planner Lee Koppleman told Caro about the low bridges on Wantagh Parkway, "The old son of a gun had made sure that buses would *never* be able to use his goddamned parkways."[7]

Histories of architecture, city planning, and public works contain many examples of physical arrangements with explicit or implicit political purposes. One can point to Baron Haussmann's broad Parisian thoroughfares, engineered at Louis Napoleon's direction to prevent any recurrence of street fighting of the kind that took place during the revolution of 1848. Or one can visit any number of grotesque concrete buildings and huge plazas constructed on university campuses in the United States during the late 1960s and early 1970s to defuse student demonstrations. Studies of industrial machines and instruments also turn up interesting political stories, including some that violate our normal expectations about why technological innovations are made in the first place. If we suppose that new technologies are introduced to achieve increased efficiency, the history of technology shows that we will sometimes be disappointed. Technological change expresses a panoply of human motives, not the least of which is the desire of some to have dominion over others even though it may require an occasional sacrifice of cost savings and some violation of the normal standard of trying to get more from less.

One poignant illustration can be found in the history of nineteenth-century industrial mechanization. At Cyrus McCormick's reaper manufacturing plant in Chicago in the middle 1880s, pneumatic molding machines, a new and largely untested innovation, were added to the foundry at an estimated cost of $500,000. The standard economic interpretation would lead us to expect that this step was taken to modernize the plant and achieve the kind of efficiencies that mechanization brings. But historian Robert Ozanne has put the development in a broader context. At the time, Cyrus McCormick II was engaged in a battle with the National Union of Iron Molders. He saw the addition of the new machines as a way to "weed out the bad element among the men," namely, the skilled workers who had organized the union local in Chicago.[8] The new machines, manned by unskilled laborers, actually produced inferior castings at a higher cost than the earlier process. After three years of use the machines were, in fact, abandoned, but by that time they had served their purpose—the destruction of the union. Thus, the story of these technical developments at the McCormick factory cannot be adequately understood outside the record of workers' attempts to organize, police repression of the labor movement in Chicago during that period, and the events surrounding the bombing at Haymarket Square. Technological history and U.S. political history were at that moment deeply intertwined.

In the examples of Moses' low bridges and McCormick's molding machines, one sees the importance of technical arrangements that precede the *use* of the things in question. It is obvious that technologies can be used in ways that enhance the power, authority, and privilege of some over others, for example, the use of television to sell a candidate. In our accustomed way of thinking technologies are seen as neutral tools that can be used well or poorly, for good, evil, or something in between. But we usually do not stop to inquire whether a given device might have been designed and built in such a way that it produces a set of consequences logically and temporally *prior to any of its professed uses*. Robert Moses' bridges, after all, were used to carry automobiles from one point to another; McCormick's machines were used to make metal castings; both technologies, however, encompassed purposes far beyond their immediate use. If our moral and political language for evaluating technology includes only categories having to do with tools and uses, if it does not include attention to the meaning of the designs and arrangements of our artifacts, then we will be blinded to much that is intellectually and practically crucial.

Because the point is most easily understood in the light of particular intentions embodied in physical form, I have so far offered illustrations that seem almost conspiratorial. But to recognize the political dimensions in the shapes of technology does not require that we look for conscious conspiracies or malicious intentions. The organized movement of handicapped people in the United States during the 1970s pointed out the countless ways in which machines, instruments, and structures of common use—buses, buildings, sidewalks, plumbing fixtures, and so forth—made it impossible for many handicapped persons to move freely about, a condition that systematically excluded them from public life. It is safe to say that designs unsuited for the handicapped arose more from long-standing neglect than from anyone's active intention. But once the issue was brought to public attention, it became evident that justice required a remedy. A whole range of artifacts have been redesigned and rebuilt to accommodate this minority.

Indeed, many of the most important examples of technologies that have political consequences are those that transcend the simple categories "intended" and "unintended" altogether. These are instances in which the very process of technical development is so thoroughly biased in a particular direction that it regularly produces results heralded as wonderful breakthroughs by some social interests and crushing setbacks by others. In such cases it is neither correct nor insightful to say, "Someone intended to do somebody else harm." Rather one must say that the technological deck has been stacked in advance to favor certain social interests and that some people were bound to receive a better hand than others.

The mechanical tomato harvester, a remarkable device perfected by researchers at the University of California from the late 1940s to the present offers an illustrative tale. The machine is able to harvest tomatoes in a single pass through a row, cutting the plants from the ground, shaking the fruit loose, and (in the newest models) sorting the tomatoes electronically into large plastic gondolas that hold up to twenty-five tons of produce headed for canning factories. To accommodate the rough motion of these harvesters in the field, agricultural researchers have bred new varieties of tomatoes that are hardier, sturdier, and less tasty than those previously grown. The harvesters replace the system of handpicking in which crews of farm workers would pass through the fields three or four times, putting ripe tomatoes in lug boxes and saving immature fruit for later harvest.[9] Studies in California indicate that the use of the machine reduces costs by approximately five to seven dollars per ton as compared to hand harvesting.[10] But the benefits are by no means equally divided in the agricultural economy. In fact, the machine in the garden has in this instance been the occasion for a thorough reshaping of social relationships involved in tomato production in rural California.

By virtue of their very size and cost of more than $50,000 each, the machines are compatible only with a highly concentrated form of tomato growing. With the introduction of this new method of harvesting, the number of tomato growers declined from approximately 4,000 in the early 1960s to about 600 in 1973, and yet there was a substantial increase in tons of tomatoes produced. By the late 1970s an estimated 32,000 jobs in the tomato industry had been eliminated as a direct consequence of mechanization.[11] Thus, a jump in productivity to the benefit of very large growers has occurred at the sacrifice of other rural agricultural communities.

The University of California's research on and development of agricultural machines such as the tomato harvester eventually became the subject of a lawsuit filed by attorneys for California Rural Legal Assistance, an organization representing a group of farm workers and other interested parties. The suit charged that university officials are spending tax monies on projects that benefit a handful of private interests to the detriment of farm workers, small farmers, consumers, and rural California generally and asks for a court injunction to stop the practice. The university denied these charges, arguing that to accept them "would require elimination of all research with any potential practical application."[12]

As far as I know, no one argued that the development of the tomato harvester was the result of a plot. Two students of the controversy, William Friedland and Amy Barton, specifically exonerate the original developers of the machine and the hard tomato from any desire to facilitate economic concentration in that industry.[13] What we see here instead is an ongoing social process in which scientific knowledge, technological invention, and corporate profit reinforce each other in deeply entrenched patterns, patterns that bear the unmistakable stamp of political and economic power. Over many decades agricultural research and development in U.S. land-grant colleges and universities has tended to favor the interests of large agribusiness concerns.[14] It is in the face of such subtly ingrained patterns that opponents of innovations such as the tomato harvester are made to seem "antitechnology" or "antiprogress." For the harvester is not merely the symbol of a social order that rewards some while punishing others; it is in a true sense an embodiment of that order.

Within a given category of technological change there are, roughly speaking, two kinds of choices that can affect the relative distribution of power, authority, and privilege in a community. Often the crucial decision is a simple "yes or no" choice— are we going to develop and adopt the thing or not? In recent years many local, national, and international disputes about technology have centered on "yes or no" judgments about such things as food additives, pesticides, the building of highways, nuclear reactors, dam projects, and proposed high-tech weapons. The fundamental choice about an antiballistic missile or supersonic transport is whether or not the thing is going to join society as a piece of its operating equipment. Reasons given for and against are frequently as important as those concerning the adoption of an important new law.

27

A second range of choices, equally critical in many instances, has to do with specific features in the design or arrangement of a technical system after the decision to go ahead with it has already been made. Even after a utility company wins permission to build a large electric power line, important controversies can remain with respect to the placement of its route and the design of its towers; even after an organization has decided to institute a system of computers, controversies can still arise with regard to the kinds of components, programs, modes of access, and other specific features the system will include. Once the mechanical tomato harvester had been developed in its basic form, a design alteration of critical social significance—the addition of electronic sorters, for example—changed the character of the machine's effects upon the balance of wealth and power in California agriculture. Some of the most interesting research on technology and politics at present focuses upon the attempt to demonstrate in a detailed, concrete fashion how seemingly innocuous design features in mass transit systems, water projects, industrial machinery, and other technologies actually mask social choices of profound significance. Historian David Noble has studied two kinds of automated machine tool systems that have different implications for the relative power of management and labor in the industries that might employ them. He has shown that although the basic electronic and mechanical components of the record/playback and numerical control systems are similar, the choice of one design over another has crucial consequences for social struggles on the shop floor. To see the matter solely in terms of cost cutting, efficiency, or the modernization of equipment is to miss a decisive element in the story.[15]

From such examples I would offer some general conclusions. These correspond to the interpretation of technologies as "forms of life" presented in the previous chapter, filling in the explicitly political dimensions of that point of view.

The things we call "technologies" are ways of building order in our world. Many technical devices and systems important in everyday life contain possibilities for many different ways of ordering human activity. Consciously or unconsciously, deliberately or inadvertently, societies choose structures for technologies that influence how people are going to work, communicate, travel, consume, and so forth over a very long time. In the processes by which structuring decisions are made, different people are situated differently and possess unequal degrees of power as well as unequal levels of awareness. By far the greatest latitude of choice exists the very first time a particular instrument, system, or technique is introduced. Because choices tend to become strongly fixed in material equipment, economic investment, and social habit, the original flexibility vanishes for all practical purposes once the initial commitments are made. In that sense technological innovations are similar to legislative acts or political foundings that establish a framework for public order that will endure over many generations. For that reason the same careful attention one would give to the rules, roles, and relationships of politics must also be given to such things as the building of highways, the creation of television networks, and the tailoring of seemingly insignificant features on new machines. The issues that divide or unite people in society are settled not only in the institutions and practices of politics proper, but also, and less obviously, in tangible arrangements of steel and concrete, wires and semiconductors, nuts and bolts.

Inherently Political Technologies

None of the arguments and examples considered thus far addresses a stronger, more troubling claim often made in writings about technology and society—the belief that some technologies are by their very nature political in a specific way. According to this view, the adoption of a given technical system unavoidably brings with it conditions for human relationships that have a distinctive political cast—for example, centralized or decentralized, egalitarian or inegalitarian, repressive or liberating. This is ultimately what is at stake in assertions such as those of Lewis Mumford that two traditions of technology, one authoritarian, the other democratic, exist side by side in Western history. In all the cases cited above the technologies are relatively flexible in design and arrangement and variable in their effects. Although one can recognize a particular result produced in a particular setting, one can also easily imagine how a roughly similar device or system might have been built or situated with very much different political consequences. The idea we must now examine and evaluate is that certain kinds of technology do not allow such flexibility, and that to choose them is to choose unalterably a particular form of political life.

A remarkably forceful statement of one version of this argument appears in Friedrich Engels's little essay "On Authority" written in 1872. Answering anarchists who believed that authority is an evil that ought to be abolished altogether, Engels launches into a panegyric for authoritarianism, maintaining, among other things, that strong authority is a necessary condition in modern industry. To advance his case in the strongest possible way, he asks his readers to imagine that the revolution has already occurred. "Supposing a social revolution dethroned the capitalists, who now exercise their authority over the production and circulation of wealth. Supposing, to adopt entirely the point of view of the anti-authoritarians, that the land and the instruments of labour had become the collective property of the workers who use them. Will authority have disappeared or will it have only changed its form?" [16]

His answer draws upon lessons from three sociotechnical systems of his day, cotton-spinning mills, railways, and ships at sea. He observes that on its way to becoming finished thread, cotton moves through a number of different operations at different locations in the factory. The workers perform a wide variety of tasks, from running the steam engine to carrying the products from one room to another. Because these tasks must be coordinated and because the timing of the work is "fixed by the authority of the steam," laborers must learn to accept a rigid discipline. They must, according to Engels, work at regular hours and agree to subordinate their individual wills to the persons in charge of factory operations. If they fail to do so, they risk the horrifying possibility that production will come to a grinding halt. Engels pulls no punches. "The automatic machinery of a big factory," he writes, "is much more despotic than the small capitalists who employ workers ever have been." [17]

Similar lessons are adduced in Engels's analysis of the necessary operating conditions for railways and ships at sea. Both require the subordination of workers to an "imperious authority" that sees to it that things run according to plan. Engels finds that far from being an idiosyncrasy of capitalist social organization, relationships of authority and subordination arise "independently of all social organization, [and] are imposed upon us together with the material conditions under which we produce and make products circulate." Again, he intends this to be stern advice to the anarchists who, according to Engels, thought it possible simply to eradicate subordination and superordination at a single stroke. All such schemes are nonsense. The roots of unavoidable authoritarianism are, he argues, deeply implanted in the human involvement with science and technology. "If man, by dint of his knowledge and inventive genius, has subdued the forces of nature, the latter avenge themselves upon him by subjecting him, insofar as he employs them, to a veritable despotism independent of all social organization."[18]

Attempts to justify strong authority on the basis of supposedly necessary conditions of technical practice have an ancient history. A pivotal theme in the *Republic* is Plato's quest to borrow the authority of *technē* and employ it by analogy to buttress his argument in favor of authority in the state. Among the illustrations he chooses, like Engels, is that of a ship on the high seas. Because large sailing vessels by their very nature need to be steered with a firm hand, sailors must yield to their captain's commands; no reasonable person believes that ships can be run democratically. Plato goes on to suggest that governing a state is rather like being captain of a ship or like practicing medicine as a physician. Much the same conditions that require central rule and decisive action in organized technical activity also create this need in government.

In Engels's argument, and arguments like it, the justification for authority is no longer made by Plato's classic analogy, but rather directly with reference to technology itself. If the basic case is as compelling as Engels believed it to be, one would expect that as a society adopted increasingly complicated technical systems as its material basis, the prospects for authoritarian ways of life would be greatly enhanced. Central control by knowledgeable people acting at the top of a rigid social hierarchy would seem increasingly prudent. In this respect his stand in "On Authority" appears to be at variance with Karl Marx's position in Volume I of *Capital*. Marx tries to show that increasing mechanization will render obsolete the hierarchical division of labor and the relationships of subordination that, in his view, were necessary during the early stages of modern manufacturing. "Modern Industry," he writes, "sweeps away by technical means the manufacturing division of labor, under which each man is bound hand and foot for life to a single detail operation. At the same time, the capitalistic form of that industry reproduces this same division of labour in a still more monstrous shape; in the factory proper, by converting the workman into a living appendage of the machine."[19] In Marx's view the conditions that will eventually dissolve the capitalist division of labor and facilitate proletarian revolution are conditions latent in industrial technology itself. The differences between Marx's position in *Capital* and Engels's in his essay raise an important question for socialism: What, after all, does modern technology make possible or necessary in political life? The theoretical tension we see here mirrors many troubles in the practice of freedom and authority that had muddied the tracks of socialist revolution.

Arguments to the effect that technologies are in some sense inherently political have been advanced in a wide variety of contexts, far too many to summarize here. My reading of such notions, however, reveals there are two basic ways of stating the case. One version claims that the adoption of a given technical system actually requires the creation and maintenance of a particular set of social conditions as the operating environment of that system. Engels's position is of this kind. A similar view is offered by a contemporary writer who holds that "if you accept nuclear power plants, you also accept a techno-scientific-industrial-military elite. Without these people in charge, you could not have nuclear power." [20] In this conception some kinds of technology require their social environments to be structured in a particular way in much the same sense that an automobile requires wheels in order to move. The thing could not exist as an effective operating entity unless certain social as well as material conditions were met. The meaning of "required" here is that of practical (rather than logical) necessity. Thus, Plato thought it a practical necessity that a ship at sea have one captain and an unquestionably obedient crew.

A second, somewhat weaker, version of the argument holds that a given kind of technology is strongly compatible with, but does not strictly require, social and political relationships of a particular stripe. Many advocates of solar energy have argued that technologies of that variety are more compatible with a democratic, egalitarian society than energy systems based on coal, oil, and nuclear power; at the same time they do not maintain that anything about solar energy requires democracy. Their case is, briefly, that solar energy is decentralizing in both a technical and political sense: technically speaking, it is vastly more reasonable to build solar systems in a disaggregated, widely distributed manner than in large-scale centralized plants; politically speaking, solar energy accommodates the attempts of individuals and local communities to manage their affairs effectively because they are dealing with systems that are more accessible, comprehensible, and controllable than huge centralized sources. In this view solar energy is desirable not only for its economic and environmental benefits, but also for the salutary institutions it is likely to permit in other areas of public life.[21]

Within both versions of the argument there is a further distinction to be made between conditions that are internal to the workings of a given technical system and those that are external to it. Engels's thesis concerns internal social relations said to be required within cotton factories and railways, for example; what such relationships mean for the condition of society at large is, for him, a separate question. In contrast, the solar advocate's belief that solar technologies are compatible with democracy pertains to the way they complement aspects of society removed from the organization of those technologies as such.

There are, then, several different directions that arguments of this kind can follow. Are the social conditions predicated said to be required by, or strongly compatible with, the workings of a given technical system? Are those conditions internal to that system or external to it (or both)? Although writings that address such questions are often unclear about what is being asserted, arguments in this general category are an important part of modern political discourse. They enter into many attempts to explain how changes in social life take place in the wake of technological innovation. More important, they are often used to buttress attempts to justify or criticize proposed courses of action involving new technology. By offering distinctly political reasons for or against the adoption of a particular technology, arguments of this kind stand apart from more commonly employed, more easily quantifiable claims about economic costs and benefits, environmental impacts, and possible risks to public health and safety that technical systems may involve. The issue here does not concern how many jobs will be created, how much income generated, how many pollutants added, or how many cancers produced. Rather, the issue has to do with ways in which choices about technology have important consequences for the form and quality of human associations.

If we examine social patterns that characterize the environments of technical systems, we find certain devices and systems almost invariably linked to specific ways of organizing power and authority. The important question is: Does this state of affairs derive from an unavoidable social response to intractable properties in the things themselves, or is it instead a pattern imposed independently by a governing body, ruling class, or some other social or cultural institution to further its own purposes?

Taking the most obvious example, the atom bomb is an inherently political artifact. As long as it exists at all, its lethal properties demand that it be controlled by a centralized, rigidly hierarchical chain of command closed to all influences that might make its workings unpredictable. The internal social system of the bomb must be authoritarian; there is no other way. The state of affairs stands as a practical necessity independent of any larger political system in which the bomb is embedded, independent of the type of regime or character of its rulers. Indeed, democratic states must try to find ways to ensure that the social structures and mentality that characterize the management of nuclear weapons do not "spin off" or "spill over" into the polity as a whole.

The bomb is, of course, a special case. The reasons very rigid relationships of authority are necessary in its immediate presence should be clear to anyone. If, however, we look for other instances in which particular varieties of technology are widely perceived to need the maintenance of a special pattern of power and authority, modern technical history contains a wealth of examples.

Alfred D. Chandler in *The Visible Hand*, a monumental study of modern business enterprise, presents impressive documentation to defend the hypothesis that the construction and day-to-day operation of many systems of production, transportation, and communication in the nineteenth and twentieth centuries require the development of particular social form—a large-scale centralized, hierarchical organization administered by highly skilled managers. Typical of Chandler's reasoning is his analysis of the growth of the railroads.[22]

> Technology made possible fast, all-weather transportation; but safe, regular, reliable movement of goods and passengers, as well as the continuing maintenance and repair of locomotives, rolling stock, and track, roadbed, stations, roundhouses, and other equipment, required the creation of a sizable administrative organization. It meant the employment of a set of managers to supervise these functional activities over an extensive geographical area; and the appointment of an administrative command of middle and top executives to monitor, evaluate, and coordinate the work of managers responsible for the day-to-day operations.

Throughout his book Chandler points to ways in which technologies used in the production and distribution of electricity, chemicals, and a wide range of industrial goods "demanded" or "required" this form of human association. "Hence, the operational requirements of railroads demanded the creation of the first administrative hierarchies in American business."[23]

Were there other conceivable ways of organizing these aggregates of people and apparatus? Chandler shows that a previously dominant social form, the small traditional family firm, simply could not handle the task in most cases. Although he does not speculate further, it is clear that he believes there is, to be realistic, very little latitude in the forms of power and authority appropriate within modern sociotechnical systems. The properties of many modern technologies—oil pipelines and refineries, for example—are such that overwhelmingly impressive economies of scale and speed are possible. If such systems are to work effectively, efficiently, quickly, and safely, certain requirements of internal social organization have to be fulfilled; the material possibilities that modern technologies make available could not be exploited otherwise. Chandler acknowledges that as one compares sociotechnical institutions of different nations, one sees "ways in which cultural attitudes, values, ideologies, political systems, and social structure affect these imperatives."[24] But the weight of argument and empirical evidence in *The Visible Hand* suggests that any significant departure from the basic pattern would be, at best, highly unlikely.

It may be that other conceivable arrangements of power and authority, for example, those of decentralized, democratic worker self-management, could prove capable of administering factories, refineries, communications systems, and railroads as well as or better than the organizations Chandler describes. Evidence from automobile assembly teams in Sweden and worker-managed plants in Yugoslavia and other countries is often presented to salvage these possibilities. Unable to settle controversies over this matter here, I merely point to what I consider to be their bone of contention. The available evidence tends to show that many large, sophisticated technological systems are in fact highly compatible with centralized, hierarchical managerial control. The interesting question, however, has to do with whether or not this pattern is in any sense a requirement of such systems, a question that is not solely empirical. The matter ultimately rests on our judgments about what steps, if any, are practically necessary in the workings of particular kinds of technology and what, if anything, such measures require of the structure of human associations. Was Plato right in saying that a ship at sea needs steering by a decisive hand and that this could only be accomplished by a single captain and an obedient crew? Is Chandler correct in saying that the properties of large-scale systems require centralized, hierarchical managerial control?

To answer such questions, we would have to examine in some detail the moral claims of practical necessity (including those advocated in the doctrines of economics) and weigh them against moral claims of other sorts, for example, the notion that it is good for sailors to participate in the command of a ship or that workers have a right to be involved in making and administering decisions in a factory. It is characteristic of societies based on large, complex technological systems, however, that moral reasons other than those of practical necessity appear increasingly obsolete, "idealistic," and irrelevant. Whatever claims one may wish to make on behalf of liberty, justice, or equality can be immediately neutralized when confronted with arguments to the effect, "Fine, but that's no way to run a railroad" (or steel mill, or airline, or communication system, and so on). Here we encounter an important quality in modern political discourse and in the way people commonly think about what measures are justified in response to the possibilities technologies make available. In many instances, to say that some technologies are inherently political is to say that certain widely accepted reasons of practical necessity—especially the need to maintain crucial technological systems as smoothly working entities—have tended to eclipse other sorts of moral and political reasoning.

One attempt to salvage the autonomy of politics from the bind of practical necessity involves the notion that conditions of human association found in the internal workings of technological systems can easily be kept separate from the polity as a whole. Americans have long rested content in the belief that arrangements of power and authority inside industrial corporations, public utilities, and the like have little bearing on public institutions, practices, and ideas at large. That "democracy stops at the factory gates" was taken as a fact of life that had nothing to do with the practice of political freedom. But can the internal politics of technology and the politics of the whole community be so easily separated? A recent study of business leaders in the United States, contemporary exemplars of

36

Chandler's "visible hand of management," found them remarkably impatient with such democratic scruples as "one man, one vote." If democracy doesn't work for the firm, the most critical institution in all of society, American executives ask, how well can it be expected to work for the government of a nation—particularly when that government attempts to interfere with the achievements of the firm? The authors of the report observe that patterns of authority that work effectively in the corporation become for businessmen "the desirable model against which to compare political and economic relationships in the rest of society."[25] While such findings are far from conclusive, they do reflect a sentiment increasingly common in the land: what dilemmas such as the energy crisis require is not a redistribution of wealth or broader public participation but, rather, stronger, centralized public and private management.

An especially vivid case in which the operational requirements of a technical system might influence the quality of public life is the debates about the risks of nuclear power. As the supply of uranium for nuclear reactors runs out, a proposed alternative fuel is the plutonium generated as a by-product in reactor cores. Well-known objections to plutonium recycling focus on its unacceptable economic costs, its risks of environmental contamination, and its dangers in regard to the international proliferation of nuclear weapons. Beyond these concerns, however, stands another less widely appreciated set of hazards—those that involve the sacrifice of civil liberties. The widespread use of plutonium as a fuel increases the chance that this toxic substance might be stolen by terrorists, organized crime, or other persons. This raises the prospect, and not a trivial one, that extraordinary measures would have to be taken to safeguard plutonium from theft and to recover it should the substance be stolen. Workers in the nuclear industry as well as ordinary citizens outside could well become subject to background security checks, covert surveillance, wiretapping, informers, and even emergency measures under martial law—all justified by the need to safeguard plutonium.

Russell W. Ayres's study of the legal ramifications of plutonium recycling concludes: "With the passage of time and the increase in the quantity of plutonium in existence will come pressure to eliminate the traditional checks the courts and legislatures place on the activities of the executive and to develop a powerful central authority better able to enforce strict safeguards." He avers that "once a quantity of plutonium had been stolen, the case for literally turning the country upside down to get it back would be overwhelming." Ayres anticipates and worries about the kinds of thinking that, I have argued, characterize inherently political technologies. It is still true that in a world in which human beings make and maintain artificial systems nothing is "required" in an absolute sense. Nevertheless, once a course of action is under way, once artifacts such as nuclear power plants have been built and put in operation, the kinds of reasoning that justify the adaptation of social life to technical requirements pop up as spontaneously as flowers in the spring. In Ayres's words, "Once recycling begins and the risks of plutonium theft become real rather than hypothetical, the case for governmental infringement of protected rights will seem compelling."[26] After a certain point, those who cannot accept the hard requirements and imperatives will be dismissed as dreamers and fools.

* * *

The two varieties of interpretation I have outlined indicate how artifacts can have political qualities. In the first instance we noticed ways in which specific features in the design or arrangement of a device or system could provide a convenient means of establishing patterns of power and authority in a given setting. Technologies of this kind have a range of flexibility in the dimensions of their material form. It is precisely because they are flexible that their consequences for society must be understood with reference to the social actors able to influence which designs and arrangements are chosen. In the second instance we examined ways in which the intractable properties of certain kinds of technology are strongly, perhaps unavoidably, linked to particular institutionalized patterns of power and authority. Here the initial choice about whether or not to adopt something is decisive in regard to its consequences. There are no alternative physical designs or arrangements that would make a significant difference; there are, furthermore, no genuine possibilities for creative intervention by different social systems—capitalist or socialist—that could change the intractability of the entity or significantly alter the quality of its political effects.

To know which variety of interpretation is applicable in a given case is often what is at stake in disputes, some of them passionate ones, about the meaning of technology for how we live. I have argued a "both/and" position here, for it seems to me that both kinds of understanding are applicable in different circumstances. Indeed, it can happen that within a particular complex of technology—a system of communication or transportation, for example—some aspects may be flexible in their possibilities for society, while other aspects may be (for better or worse) completely intractable. The two varieties of interpretation I have examined here can overlap and intersect at many points.

These are, of course, issues on which people can disagree. Thus, some proponents of energy from renewable resources now believe they have at last discovered a set of intrinsically democratic, egalitarian, communitarian technologies. In my best estimation, however, the social consequences of building renewable energy systems will surely depend on the specific configurations of both hardware and the social institutions created to bring that energy to us. It may be that we will find ways to turn this silk purse into a sow's ear. By comparison, advocates of the further development of nuclear power seem to believe that they are working on a rather flexible technology whose adverse social effects can be fixed by changing the design parameters of reactors and nuclear waste disposal systems. For reasons indicated above, I believe them to be dead wrong in that faith. Yes, we may be able to manage some of the "risks" to public health and safety that nuclear power brings. But as society adapts to the more dangerous and apparently indelible features of nuclear power, what will be the long-range toll in human freedom?

My belief that we ought to attend more closely to technical objects themselves is not to say that we can ignore the contexts in which those objects are situated. A ship at sea may well require, as Plato and Engels insisted, a single captain and obedient crew. But a ship out of service, parked at the dock, needs only a caretaker. To understand which technologies and which contexts are important to us, and why, is an enterprise that must involve both the study of specific technical systems and their history as well as a thorough grasp of the concepts and controversies of political theory. In our times people are often willing to make drastic changes in the way they live to accommodate technological innovation while at the same time resisting similar kinds of changes justified on political grounds. If for no other reason than that, it is important for us to achieve a clearer view of these matters than has been our habit so far.

3
TECHNĒ AND POLITEIA

To ACHIEVE a political understanding of technology requires that we examine the realm of tools and instruments from a fresh point of view. We have already begun to recognize some of the ways in which conditions of power, authority, freedom, and social justice are deeply embedded in technical structures. From this standpoint no part of modern technology can be judged neutral a priori. All varieties of hardware and their corresponding forms of social life must be scrutinized to see whether they are friendly or unfriendly to the idea of a just society.

But what does the study of politics have to contribute to our thinking about the realm of instrumental things? Where can one turn to look for a political theory of technology?

A Classic Analogy

In THE previous chapter I noted that rooted in Western political thought is a powerful analogy linking the practice of technology to that of politics. In his *Republic, Laws, Statesman,* and other dialogues, Plato asserts that statecraft is a *technē,* one of the practical arts. Much like architecture, weaving, shipbuilding, and other arts and crafts, politics is a field of practice that has its own distinctive knowledge, its own special skills. As we have seen, one purpose of Plato's argument was to discredit those who believed that the affairs of public life could be left to mere amateurs, the democratic masses. But beyond that it is clear he thought the art of politics could be useful in the same way as any other *technē,* that it could produce well-crafted works of lasting value.

The works he had in mind were good constitutions, supremely well-crafted products of political architecture. *Politeia*, the title of the *Republic* in Greek, means the constitution of a polis, the proper order of human relationships within a city-state. The dialogue describes and justifies what Plato holds to be the institutional arrangements appropriate to the best *politeia*. He returns to this theme in the *Laws*, a discussion of the "second best" constitution, comparing his work to that of a well-established craft. "The shipwright, you know, begins his work by laying down the keel of the vessel and indicating her outlines, and I feel myself to be doing the same thing in my attempt to present you with outlines of human lives. . . . I am really laying the keels of the vessels by due consideration of the question by what means or manner of life we shall make our voyage over the sea of time to the best purpose."[1] There is evidence that Plato actually sought to realize his skills as a designer/builder of political societies. He traveled from Athens to live at the court of Dionysius the Elder, tyrant of Syracuse, hoping to transform his host into a genuine philosopher-king, a person willing to apply the true principles of political *technē*. The attempt did not succeed.

In Plato's interpretation the analogy between technology and politics works in one direction only; *technē* serves as a model for politics and not the other way around. Although respectful of the power of the material arts, he remained deeply suspicious of them. Thus, in the *Laws* he excludes craftsmen from positions of citizenship, explaining that they already have an art that requires their full attention. At the same time he forbids citizens to engage in any material craft whatsoever because citizenship makes full demands on them. Plato's discomfort with technology has remained characteristic of moral and political philosophers to this day. Most of them have politely ignored the substance of technical life, hoping perhaps that it would remain segregated in a narrowly defined corner of human life. Evidently, it did not occur to anyone that Plato's pregnant analogy would at some point qualify in reverse, that *technē* itself might become *politeia*, that technical forms of life might in themselves play a powerful role in shaping society. When that finally happened, political theory would find itself totally unprepared.

The one-sided comparison of technical and political creativity appears again in modern political thought. Writing in *The Social Contract*, Jean-Jacques Rousseau employs a mechanical metaphor to illuminate the art of constitution making. "A prince," he says, "has only to follow a model which the lawgiver provides. The lawgiver is the engineer who invents the machine; the prince is merely the mechanic who sets it up and operates it."[2] At another point in the book, Rousseau compares the work of the lawgiver to that of an architect. With a frustrated ambition reminiscent of Plato's, Rousseau offered himself as a political engineer or architect of exactly this kind, writing treatises on the constitutions of Corsica and Poland in the hope that his ideas might influence the founding of new states.

A practical opportunity of exactly that kind later became available to the founders of modern nation states, among them the leaders of the American Revolution. From the earliest rumblings of rebellion in the seventeenth century to the adoption of the U.S. Constitution in 1787, the nation was alive with disputes about the application of political principles to the design of public institutions. Once again the ancient analogy between politics and technology became an expressive idea. Taking what they found useful from previous history and existing theories, thinkers like Madison, Hamilton, Adams, and Jefferson tried to devise a "science of politics," a science specifically aimed at providing knowledge for a collective act of architectonic skill. Thus, in *The Federalist Papers*, to take one example, we find a sustained discussion of how to move from abstract political notions such as power, liberty, and public good to their tangible manifestations in the divisions, functions, powers, relationships, and limits of the Constitution. "The science of politics," Hamilton explains in "Federalist No. 9," "like most other sciences, has received great improvement. The efficacy of various principles is now well understood, which were either not known at all, or imperfectly known to the ancients. The regular distribution of power into distinct departments; the introduction of legislative balances and checks; the institution of courts composed of judges holding their offices during good behavior; the representation of the people in the legislature by deputies of their own election: these are wholly new discoveries, or have made their principal progress towards perfection in modern times." Metaphors from eighteenth-century science and mechanical invention—for example, "the ENLARGEMENT of the ORBIT within which such systems are to revolve"[3] and references to the idea of checks and balances—pervade *The Federalist Papers* and indicate the extent to which its writers saw the founding as the creation of an ingenious political/mechanical device.

But even as the eighteenth century was reviving the comparison between technology and politics, even as philosopher statesmen were restoring the *technē* of constitution making, another extremely powerful mode of institutionalization was taking shape in the United States and Europe. The industrial revolution with its distinctive ways of arranging people, machines, and materials for production very soon began to compete with strictly political institutions for power, authority, and the loyalties of men and women. Writing in 1781 in his *Notes on Virginia*, Thomas Jefferson noted the new force abroad in the world and commented upon its probable meaning for political society. The system of manufacturing emerging at the time would, he argued, be incompatible with the life of a stable, virtuous republic. Manufacturing would create a thoroughly dependent rather than a self-sufficient populace. "Dependence," he warned, "begets subservience and venality, suffocates the germ of virtue, and prepares fit tools for the designs of ambition." In his view the industrial mode of production threatened "the manners and spirit of a people which preserve a republic in vigor. A degeneracy in these is a canker which soon eats to the heart of its laws and constitution."[4] For that reason he advised, in this book at least, that Americans agree to leave the workshops in Europe.

Abundance and Freedom

JEFFERSON'S PLEA echoes a belief common in writings of ancient Greece and Rome that civic virtue and material prosperity are antithetical. Human nature, according to this view, is easily corrupted by wealth. The indolent, pleasure-seeking habits of luxurious living tend to subvert qualities of frugality, self-restraint, and self-sacrifice needed to maintain a free society. By implication, any society that wishes to maintain civic virtue ought to approach technical innovation and economic growth with the utmost caution. At the time of the founding of the American republic, the country did not depend upon high levels of material production and consumption. In fact, during political discussions of the 1770s and 1780s, the quest for material wealth was sometimes mentioned as a danger, a source of corruption. A speaker before the Continental Congress in 1775 called upon the citizenry to "banish the syren LUXURY with all her train of fascinating pleasures, idle dissipation, and expensive amusements from our borders," and to institute "honest industry, sober frugality, simplicity of manners, and plain hospitality and christian benevolence."[5]

There are signs that a desire to shape industrial development to accord with the ideals of the republican political tradition continued to interest some Americans well into the 1830s. Attempts to include elements of a republican community in the building of the factory town in Lowell, Massachusetts, show this impulse at work.[6] But these efforts were neither prominent in the economic patterns then taking shape nor successful in their own right. In the 1840s and decades since the notion that industrial development might be shaped or limited by republican virtues dropped out of common discourse, echoed only in the woeful lamentations of Henry David Thoreau, Henry Adams, Lewis Mumford, Paul Goodman, and a host of others now flippantly dismissed as "romantics" and "pastoralists."

In fact, the republican tradition of political thought had long since made its peace with the primary carrier of technical change, entrepreneurial capitalism. Moral and political thinkers from Machiavelli to Montesquieu and Adam Smith had argued, contrary to the ancient wisdom, that the pursuit of economic advantage is actually a civilizing, moderating influence in society, the very basis of stable government. Rather than engage the fierce passion for glory that often leads to conflict, it is better, so the argument goes, to convince people to pursue their self-interest, an interest that inclines them toward rational behavior.[7] The framers of the American Constitution were, by and large, convinced of the wisdom of this formula. They expected that Americans would act in a self-interested manner, employing whatever instruments they needed to generate wealth. The competition of these interests in society would, they believed, provide a check upon the concentration of power in the hands of any one faction. Thus, in one important sense republicanism and capitalism were fully reconciled at the time of the founding.

By the middle of the nineteenth century this point of view had been strongly augmented by another idea, one that to this day forms the basic self-image of Americans—a notion that equates abundance and freedom. The country was rich in land and resources; people liberated from the social hierarchies and status definitions of traditional societies were given the opportunity to exploit that material bounty in whatever ways they could muster. In this context new technologies were seen as an undeniable blessing because they enabled the treasures to be extracted more quickly, because they vastly increased the product of labor. Factories, railroads, steamboats, telegraphs, and the like were greeted as the very essence of democratic freedom for the ways they rendered, as one mid–nineteenth-century writer explained, "the conveniences and elegancies of life accessible to the many instead of the few."[8]

American society encouraged people to be self-determining, to pursue their own economic goals. That policy would work, it was commonly believed, only if there were a surplus that guaranteed enough to go around. Class conflict, the scourge of democracy in the ancient world, could be avoided in the United States because the inequalities present in society would not matter very much. Material abundance would make it possible for everybody to have enough to be perfectly happy. Eventually, Americans took this notion to be a generally applicable theory: economic enterprise driven by the engine of technical improvement was the very essence of human freedom. Franklin D. Roosevelt reportedly remarked that if he could put one American book in the hands of every Russian, it would be the Sears, Roebuck catalogue.

In this way of looking at things the form of the technology you adopt does not matter. If you have cornucopia in your grasp, you do not worry about its shape. Insofar as it is a powerful thing, more power to it. Anything that history, literature, philosophy, or long-standing traditions might have to suggest about the prudence one ought to employ in the shaping of new institutions can be thrown in the trash bin. Describing the industrial revolution in Britain, historian Karl Polanyi drew an accurate picture of this attitude. "Fired by an emotional faith in spontaneity, the common-sense attitude toward change was discarded in favor of a mystical readiness to accept the social consequences of economic improvement, whatever they might be. The elementary truths of political science and statecraft were first discarded, then forgotten. It should need no elaboration that a process of undirected change, the pace of which is deemed too fast, should be slowed down, if possible, so as to safeguard the welfare of the community. Such household truths of traditional statesmanship, often merely reflecting the teachings of a social philosophy inherited from the ancients, were in the nineteenth century erased from the thoughts of the educated by the corrosive of a crude utilitarianism combined with an uncritical reliance on the alleged self-healing virtues of unconscious growth."[9] Indeed, by the late nineteenth century, an impressive array of scientific discoveries, technical inventions, and industrial innovations seemed to make the mastery of nature an accomplished fact rather than an idle dream. Many took this as a sign that all ancient wisdom, like all the old-fashioned machines and techniques, had simply been rendered obsolete. As one chronicler of the new technology wrote in *Scientific American*: "The speculative philosophy of the past is but a too empty consolation for short-lived, busy man, and, seeing with the eye of science the possibilities of matter, he has touched it with the divine breath of thought and made a new world."[10]

According to this view, everything one might desire of the relationship between expanding industrial technology and the building of a good society will happen automatically. All that is necessary is to make sure the machinery is up to date, well maintained, and well oiled. The only truly urgent questions that remain are ones of technical and economic efficiency. For unless a society keeps pace with the most efficient means available anywhere in the world, it will lag behind its competitors, a precondition of cultural decline.

A fascination with efficiency is a venerable tradition in American life. It is announced early on, for example, in Benjamin Franklin's maxims that economizing on time, effort, and money is a virtue. With the advance of industrialism in the late nineteenth and early twentieth centuries, this concern grew to something of an obsession among the well-educated in the United States. Understood to be a criterion applicable to personal and social life as well as to mechanical and economic systems, efficiency was upheld as a goal supremely valuable in its own right, one strongly linked to the progress of science, the growth of industry, the rise of professionalism, and the conservation of natural resources. During the Progressive Era the rule of efficient, well-trained professionals was upheld as a way of sanitizing government of the corruption of party machines and eliminating the influence of selfish interest groups. An eagerness to define important public issues as questions of efficiency has continued to be a favorite strategy in American politics over many decades; adherence to this norm has been (and still is) welcomed as the best way to achieve the ends of democracy without having to deal with democracy as a living political process. Demonstrat-

46

ing the efficiency of a course of action conveys an aura of scientific truth, social consensus, and compelling moral urgency. And Americans do not even worry much about the specific content of numerators and denominators used in efficiency measurements. As long as they are getting more for less, all is well.[11]

The Technical Constitution of Society

WITH THE PASSAGE of time the cornucopia of modern industrial production began to generate some distinctive institutional patterns. Today we can examine the interconnected systems of manufacturing, communications, transportation, and the like that have arisen during the past two centuries and appreciate how they form de facto a constitution of sorts, the constitution of a sociotechnical order. This way of arranging people and things, of course, did not develop as the result of the application of any particular plan or political theory. It grew gradually and in separate increments, invention by invention, industry by industry, engineering project by engineering project, system by system. From a contemporary vantage point, nevertheless, one can notice some of its characteristics and begin to see how they embody answers to age-old political questions—questions about membership, power, authority, order, freedom, and justice. Several of the characteristics that matter in this way of seeing things—characteristics that would certainly have interested Plato, Rousseau, Madison, Hamilton, and Jefferson—can be summarized as follows.

First is the ability of technologies of transportation and communication to facilitate control over events from a single center or small number of centers. Largely unchecked by effective countervailing influences, there has been an extraordinary centralization of social control in large business corporations, bureaucracies, and the military. It has seemed an expedient, rational way of doing things. Without anyone having explicitly chosen it, dependency upon highly centralized organizations has gradually become a dominant social form.

Second is a tendency for new devices and techniques to increase the most efficient or effective size of organized human associations. Over the past century more and more people have found themselves living and working within technology-based institutions that previous generations would have called gigantic. Justified by impressive economies of scale and, economies or not, always an expression of the power that accrues to very large organizations, this gigantism has become an accustomed feature in the material and social settings of everyday life.

Third is the way in which the rational arrangement of sociotechnical systems has tended to produce its own distinctive forms of hierarchical authority. Legitimized by the felt need to do things in what seems to be the most efficient, productive way, human roles and relationships are structured in rule-guided patterns that involve taking orders and giving orders along an elaborate chain of command. Thus, far from being a place of democratic freedom, the reality of the workplace tends to be undisguisedly authoritarian. At higher levels in the hierarchy, of course, professionals claim their special authority and relative freedom by virtue of their command of scientific and technical expertise. At the point in history in which forms of hierarchy based on religion and tradition had begun to crumble, the need to build and maintain technical systems offered a way to restore pyramidal social relations. It was a godsend for inequality.

Fourth is the tendency of large, centralized, hierarchically arranged sociotechnical entities to crowd out and eliminate other varieties of human activity. Hence, industrial techniques eclipsed craftwork; technologies of modern agribusiness made small-scale farming all but impossible; high-speed transportation crowded out slower means of getting about. It is not merely that useful devices and techniques of earlier periods have been rendered extinct, but also that patterns of social existence and individual experience that employed these tools have vanished as living realities.

Fifth are the various ways that large sociotechnical organizations exercise power to control the social and political influences that ostensibly control them. Human needs, markets, and political institutions that might regulate technology-based systems are often subject to manipulation by those very systems. Thus, to take one example, psychologically sophisticated techniques of advertising have become a customary way of altering people's ends to suit the structure of available means, a practice that now affects political campaigns no less than campaigns to sell underarm deodorant or Coca-Cola (with similar results).

There are many other characteristics of today's technological systems that can accurately be read as political phenomena. And it is certainly true that there are factors other than technology that strongly influence the developments I have mentioned. But it is important to note that as our society adopts one sociotechnical system after another it answers some of the most important questions that political philosophers have ever asked about the proper order of human affairs. Should power be centralized or dispersed? What is the best size for units of social organization? What constitutes justifiable authority in human associations? Does a free society depend upon social uniformity or diversity? What are appropriate structures and processes of public deliberation and decision making? For the past century or longer our responses to such questions have often been instrumental ones, expressed in an instrumental language of efficiency and productivity, physically embodied in human/machine systems that seem to be nothing more than ways of providing goods and services.

If we compare the process through which today's sociotechnical constitution evolved to the process employed by the framers of the U.S. Constitution, the contrast is striking. Clearly, the founding fathers considered all of the crucial questions in classical political thought. When they included a particular feature in the structure of government, it was because they had studied the matter, debated it, and deliberately chosen the result. They understood that if all went well the structures they were building would last a good long time. Because the rules, roles, and relationships they agreed upon would shape the future life of a whole nation, the framers acknowledged a special responsibility—a responsibility of wise political craftsmanship. To realize this responsibility required a depth of knowledge about political institutions and sensitivity to human motives altogether rare in human history. The results of their work include two centuries of relatively stable government in the United States, a sign that they practiced their craft well.

Of course, there are founding fathers of our sociotechnical constitution as well—the inventors, entrepreneurs, financiers, engineers, and managers who have fashioned the material and social dimensions of new technologies. Some of their names are well known to the public—Thomas Edison, Henry Ford, J. P. Morgan, John D. Rockefeller, Alfred P. Sloan, Thomas Watson, and the like. The names of such figures as Theodore Vail, Samuel Insull, and William Mullholland are not household words, but their accomplishments as builders of technological infrastructures are equally impressive. In one sense the founders of technological systems are no strangers to politics; many of them have engaged in fierce political struggles to realize their aims. William Mullholland's ruthless machinations to bring Owens Valley water to Los Angeles' desert climate is a classic case in point.[12] But the qualities of political wisdom we find in the founders of the U.S. Constitution are, by and large, missing in those who design, engineer, and promote vast systems. Here the founding fathers have been concerned with such matters as the quest for profits, organizational control, and the pleasures of innovation. They have seldom been interested in the significance of their work on the overall structure of society or its justice.

For those who have embraced the formula of freedom through abundance, however, questions about the proper order of society do not matter very much. Over many decades technological optimists have been sustained by the belief that whatever happened to be created in the sphere of material/instrumental culture would certainly be compatible with freedom, democracy, and social justice. This amounts to a conviction that all technology—whatever its size, shape, or complexion—is inherently liberating. For reasons noted in the previous chapter, that is a very peculiar faith indeed.

It is true that on occasion agencies of the modern state have attempted to "regulate" business enterprises and technological applications of various kinds. On balance, however, the extent of that regulation has been modest. In the United States absolute monopolies are sometimes outlawed only to be replaced by enormous semimonopolies no less powerful in their ability to influence social outcomes. The history of regulation shows abundant instances in which the rules and procedures that govern production or trade were actually demanded or later captured by the industries they supposedly regulate. In general, the rule of thumb has been if a business makes goods and services widely available, at low cost with due regard for public health and safety, and with a reasonable return on investment, the republic is well served.[13]

In recent times the idea of recognizing limits upon the growth of certain technologies has experienced something of a revival. Many people are prepared to entertain the notion of limiting a given technology if:

1. Its application threatens public health or safety
2. Its use threatens to exhaust some vital resource
3. It degrades the quality of the environment (air, land, and water)

4. It threatens natural species and wilderness areas that ought
   to be preserved
5. Its application causes social stresses and strains of an exaggerated kind.

Along with ongoing discussions about ways to sustain economic growth, national competitiveness, and prosperity, these
are the only matters of technology assessment that the general
public, decision makers, and academicians are prepared to take
seriously.

While such concerns are valid, they severely restrict the range
of moral and political criteria that are permissible in public deliberations about technological change. Several years ago I tried
to register my discomfort on this score with some colleagues in
computer science and sociology who were doing a study of the
then-novel systems of electronic funds transfer (EFT). They had
concluded that such systems contained the potential for re-
distributing financial power in the world of banking. Electronic
money would make possible a shift of power from smaller banks
to large national and international financial institutions. Beyond
that it appeared such systems posed serious problems about data
protection and individual privacy. They asked me to suggest an
effective way of presenting the possible dangers of this development to their audience of scholars and policy makers. I recommended that their research try to show that under conditions of
heavy, continued exposure, EFT causes cancer in laboratory animals. Surely, that finding would be cause for concern. My ironic
suggestion acknowledged what I take to be the central characteristic of socially acceptable criticism of technology in our
time. Unless one can demonstrate conclusively that a particular
technical practice will generate some physically evident catastrophe—cancer, birth defects, destruction of the ozone layer, or
some other—one might as well remain silent.

The conversation about technology and society has continued to a point at which an obvious question needs to be addressed: Are there no shared ends that matter to us any longer
other than the desire to be affluent while avoiding the risk of
cancer? It may be that the answer is no. The prevailing consensus
seems to be that people love a life of high consumption, tremble
at the thought that it might end, and are displeased about having
to clean up the messes that modern technologies sometimes
bring. To argue a moral position convincingly these days requires that one speak to (and not depart from) people's love of

material well-being, their fascination with efficiency, or their fear of death. The moral sentiments that hold force can be arrayed on a spectrum ranging from Adam Smith to Frederick W. Taylor to Thomas Hobbes. I do not wish to deny the validity of these sentiments, only to point out that they represent an extremely narrow mindset. Concerns about particular technological hazards are sometimes the beginning of a much broader political awareness. But for the most part we continue to disregard a problem that has been brewing since the earliest days of the industrial revolution—whether our society can establish forms and limits for technological change, forms and limits that derive from a positively articulated idea of what society ought to be.

As a way of beginning that project, I would suggest a simple heuristic exercise. Let us suppose that every political philosophy in a given time implies a technology or set of technologies in a particular pattern for its realization. And let us recognize that every technology of significance to us implies a set of political commitments that can be identified if one looks carefully enough. What appear to be merely instrumental choices are better seen as choices about the form of social and political life a society builds, choices about the kinds of people we want to become. Plato's metaphor, especially his reference to the shipwright, is one that an age of high technology ought to ponder carefully: we ought to lay out the keels of our vessels with due consideration to what means or manner of life best serves our purpose in our voyage over the sea of time. The vessels that matter now are such things as communications systems, transit systems, energy supply and distribution systems, information networks, household instruments, biomedical technologies, and of course systems of industrial and agricultural production. Just as Plato and Aristotle posed the question, What is the best form of political society? so also an age of high technology ought to ask, What forms of technology are compatible with the kind of society we want to build?

Answers to that question often appear as subliminal themes or concealed agendas in policy discussions that seem to be about productivity, efficiency, and economic growth. A perfect set of examples can be found among the dozens of sophisticated energy studies conducted during the 1970s in response to what was then called "the energy crisis." A careful reader can survey the various reports and interpret the political and social structures their analyses and recommendations imply.[14] Would it be nuclear power administered by a benign priesthood of scientists? Would it be coal and oil brought to you by large, multinational corporations? Would it be synthetic fuels subsidized and administered by the state? Or would it be the soft energy path brought to you by you and your neighbors?

Whatever one's position might be, the prevailing consensus required that all parties base their arguments on a familiar premise: efficiency. Regardless of how a particular energy solution would affect the distribution of wealth and social power, the case for or against it had to be stated as a practical necessity deriving from demonstrable conditions of technical or economic efficiency. As the Ford Foundation's Nuclear Energy Policy Study Group explained: "When analyzing energy, one must first decide whether ordinary rules of economics can be applied." The group decided that, yes, energy should be considered "an economic variable, rather than something requiring special analysis."[15] After that decision had been made, of course, the rest was simply a matter of putting Btus or kilowatt-hours in the numerator and dollars in the denominator and worshipping the resulting ratio as gospel.

Even those who held unorthodox viewpoints in this debate found it necessary to uphold the supreme importance of this criterion. Thus, Amory B. Lovins, a leading proponent of soft energy paths, wrote of his method: "While not under the illusion that facts are separable from values, I have tried . . . to separate my personal preferences from my analytic assumptions and to rely not on modes of discourse that might be viewed as overtly ideological, but rather on classicial arguments of economic and engineering efficiency (which are only tacitly ideological)."[16] To Lovins's credit, he consistently argued that the social consequences of energy choices were, in the last analysis, the most important aspect of energy policy making. In his widely read *Soft Energy Paths*, Lovins called attention to "centrism, vulnerability, technocracy, repression, alienation" and other grave problems that afflict conventional energy solutions. Lovins compares "two energy paths that are distinguished by their antithetical social implications." He notes that basing energy decisions on social criteria may appear to involve a "heroic decision," that is, "doing something the more expensive way because it is desirable on other more important grounds than internal cost."

But Lovins is careful not to appeal to his readers' sense of courage or altruism. "Surprisingly," he writes, "a heroic decision does not seem to be necessary in this case, because the energy system that seems socially more attractive is also cheaper and easier."[17] But what if the analysis had shown the contrary? Would Lovins have been prepared to give up the social advantages believed to exist along the soft energy path? Would he have accepted "centrism, vulnerability, technocracy, repression, alienation," and the like? Here Lovins yielded ground that in recent history has again and again been abandoned as lost territory. It raises the question of whether even the best intentioned, best qualified analysts in technological decision making are anything more than mere efficiency worshippers.

Much the same strategy often appears in the arguments of those who favor democratic self-management, decentralization, and human-scale technology. As Paul Goodman once noted, "Now, if lecturing at a college, I happen to mention that some function of society which is highly centralized could be much decentralized without loss of efficiency, or perhaps with a gain in efficiency, at once the students want to talk about nothing else."[18] That approach is, indeed, one way of catching people's attention; if you can get away with it, it is certainly a most convincing kind of argument. Because the idea of efficiency attracts a wide consensus, it is sometimes used as a conceptual Trojan horse by those who have more challenging political agendas they hope to smuggle in. But victories won in this way are in other respects great losses. For they affirm in our words and in our methodologies that there are certain human ends that no longer dare be spoken in public. Lingering in that stuffy Trojan horse too long, even soldiers of virtue eventually suffocate.

Regimes of Instrumentality

IN OUR TIME *technē* has at last become *politeia*—our instruments are institutions in the making. The idea that a society might try to guide its sociotechnical development according to self-conscious, critically evaluated standards of form and limit can no longer be considered a "heroic decision"; it is simply good sense. Because technological innovation is inextricably linked to processes of social reconstruction, any society that hopes to control its own structural evolution must confront each significant set of technological possibilities with scrupulous care.

Applied in this setting, political theory can help reveal strategic decisions in technological design. From its perspective, each significant area of technical/functional organization in modern society can be seen as a kind of regime, a regime of instrumentality, under which we are obliged to live. Thus, there are a number of regimes of mass production, each with a structure that may be interpreted as a technopolitical phenomenon. There are a number of regimes in energy production and distribution, in petroleum, coal, hydroelectricity, nuclear power, etc., each with a form that can be scrutinized for the politics of its structural properties. There is, of course, the regime of broadcast television and that of the automobile. If we were to identify and characterize all of the sociotechnical systems in our society, all of our regimes of instrumentality and their complex interconnections, we would have a clear picture of the second constitution I mentioned earlier, one that stands parallel to and occasionally overlaps the constitution of political society as such.

The important task becomes, therefore, not that of studying the "effects" and "impacts" of technical change, but one of evaluating the material and social infrastructures specific technologies create for our life's activity. We should try to imagine and seek to build technical regimes compatible with freedom, social justice, and other key political ends. Insofar as the possibilities present in a given technology allow it, the thing ought to be designed in both its hardware and social components to accord with a deliberately articulated, widely shared notion of a society worthy of our care and loyalty. If it is clear that the social contract implicitly created by implementing a particular generic variety of technology is incompatible with the kind of society we deliberately choose—that is, if we are confronted with an inherently political technology of an unfriendly sort—then that kind of device or system ought to be excluded from society altogether.

What I am suggesting is a process of technological change disciplined by the political wisdom of democracy. It would require qualities of judiciousness in the populace that have rarely been applied to the judgment of instrumental/functional affairs. It would, presumably, produce results sometimes much different from those recommended by the rules of technical and economic efficiency. Other social and political norms, articulated by a democratic process, would gain renewed prominence. Faced with any proposal for a new technological system, citizens or their representatives would examine the social contract implied by building that system in a particular form. They would ask, How well do the proposed conditions match our best sense of who we are and what we want this society to be? Who gains and who loses power in the proposed change? Are the conditions produced by the change compatible with equality, social justice, and the common good?. To nurture this process would require building institutions in which the claims of technical expertise and those of a democratic citizenry would regularly meet face to face. Here the crucial deliberations would take place, revealing the substance of each person's arguments and interests. The heretofore concealed importance of technological choices would become a matter for explicit study and debate.

There are any number of ways in which the structural features of instrumental regimes might become a focus for democratic decision making. Technologies introduced in the workplace are the ones most often mentioned; in such cases it is usually fairly clear, as in developments in automation and robotization, whose interests are immediately helped and harmed in building a new system. There are, however, a wide variety of areas in which the political complexion of technological systems could fruitfully be explored. The following illustration is one in which there is no "crisis" or obvious social problem at hand, but in which the shape of an evolving system has interesting political dimensions.

The field of solar electricity, photovoltaic energy, is one in which the crucial choices are, to a certain extent, still open for discussion. We can expect to see events unfold in our lifetimes with outcomes that could have many different dimensions. If solar cells become feasible to mass produce, if their price in installed systems comes down to a reasonable level, solar electricity could make a contribution to our society's aggregate energy needs. If the day arrives that photovoltaic systems are technically and economically feasible (and many who work with solar electric prototypes believe they will be), there will be—at least in principle—a choice about how society will structure these systems. One could, for example, build centralized photovoltaic farms that hook directly into the existing electrical grid like any other form of centrally generated electric power. It might also be possible to produce a great number of stand-alone systems placed on the rooftops of homes, schools, factories, and the like. Or one could design and build medium-sized ensembles, perhaps at a neighborhood level. When it comes time to choose which model of photovoltaic development our society will have, a number of implicit questions will somehow be answered. How large should such systems be? How many will be built? Who should own them? How should they be managed? Should they be fully automatic? Or should the producer/consumer of solar power be actively involved in activities of load management?

All of these are questions about the shape of a new regime of instrumentality. What kind of regime do we wish to build? What material and social structure should it have? In light of the patterns of technological development I mentioned earlier—patterns of centralization, gigantism, hierarchical authority, and so forth—perhaps it would be desirable to choose a model of photovoltaic development based on a more flexible, more democratic principle. Here is an opportunity to extend responsibility and control to a greater number of people, an opportunity to create diversity rather than uniformity in our sociotechnical constitution. Is this not an opportunity we should welcome and seek to realize? Suggesting this, I would not ask anyone to make exorbitant economic sacrifices. But rather than pursue the lemming-like course of choosing only that system design which provides the least expensive kilowatt, perhaps we ought to consider which system might play the more positive role in the technical infrastructure of freedom.

It goes without saying that the agencies now actually developing photovoltaics have no such questions in mind. Government-subsidized research, such as it is, focuses upon finding the most efficient and effective form of solar electricity and seeing it marketed in the "private sector."[19] Huge multinational petroleum corporations have bought up the smaller companies at work in this field; their motive seems to be desire to control the configuration of whatever mix of energy sources and technologies we will eventually have. As we have done so often in the past, our society has, in effect, delegated decision-making power to those whose plans are narrowly self-interested. One can predict, therefore, that when photovoltaic systems are introduced they will carry the same qualities of institutional and physical centralization that characterize so many modern technologies.

A crucial failure in modern political thought and political practice has been an inability or unwillingness even to begin the project I am suggesting here: the critical evaluation and control of our society's technical constitution. The silence of liberalism on this issue is matched by an equally obvious neglect in Marxist theory. Both persuasions have enthusiastically sought freedom in sheer material plenitude, welcoming whatever technological means (or monstrosities) seemed to produce abundance the fastest. It is, however, a serious mistake to construct one sociotechnical system after another in the blind faith that each will turn out to be politically benign. Many crucial choices about the forms and limits of our regimes of instrumentality must be enforced at the founding, at the genesis of each new technology. It is here that our best purposes must be heard.

II

TECHNOLOGY: REFORM AND REVOLUTION

4
BUILDING THE BETTER MOUSETRAP

To observe its custom of advertising outrageously trendy, exorbitantly expensive Christmas gifts, the 1977 Neiman-Marcus catalogue announced the sale of "His and Her Urban Windmills." Priced at $16,000 each, the devices would, according to the department store, enable customers to "enjoy today's electrical appliances and gadgets without overtaxing public power supplies, family utility bills, or tempers." The ad assured prospective buyers that the windmills would be nonpolluting, noiseless, and environmentally safe. For the woman living in a city with an average wind speed of twelve miles per hour, the mill "could generate more than enough wattage to brew her morning coffee, Benedict an egg, heat her hair rollers, soothe her psyche with stereo, and give her bronze beauty while she relaxes under the sun lamp." Other uses cited for the mill's energy include blending daiquiris and supplying power for junior's rock band.

For anyone who had followed the rise of the appropriate technology movement during the middle 1970s, the ad had an ominous ring. Would the quest for ecologically sound, small-scale, humane technologies turn out to be a mere fad? Corrupted by Madison Avenue, commercialization, and bureaucratic co-optation, movements for social change in late twentieth-century America have often ended up as fashion trends. In a matter of weeks the radical thrust of a new idea can be absorbed into the glossy, ephemeral surfaces of the postindustrial marketplace. Would appropriate technology meet the same dreary fate?

The notion of an "appropriate," "intermediate," or "alternative" technology was first proposed in the mid-1960s as a way

of addressing the economic, technical, and social problems of
Third World countries. British economist E. F. Schumacher and
others with similar views tried to develop methods of economic
improvement in harmony with the existing conditions of traditional societies. Technical development would take heed of the
skill levels of a population, natural resources available indige-
nously, and pressing social needs defined by the people themselves. Rather than introduce the biggest and best technologies
of Western industrialized countries, the idea was to devise a collection of small-scale agricultural and industrial techniques, for
example, Schumacher's one-person egg carton factory, that
would solve immediate problems without causing serious cultural disruption.[1]

Beginning in the early 1970s the concept of appropriate technology was applied to the problems of advanced industrial
societies as well. Social activists argued that pollution, environmental damage, spiraling energy costs, resource depletion,
alienation, and other gnawing social ills could be remedied if the
right kinds of technology were widely used. "Wisdom demands
a new orientation of science and technology towards the organic, the gentle, the non-violent, the elegant and beautiful,"
Schumacher argued in *Small is Beautiful*, one of the most widely
read books of social philosophy of the decade.[2] A number of organizations took up the challenge, seeking to overcome the destructive practices of modern civilization. The New Alchemy
Institute, Farallones Institute, Intermediate Technology Development Group, countless solar energy companies, and ecologically oriented research institutes in the United States, Canada,
and Europe set about exploring a variety of technological alternatives.[3] In California Governor Jerry Brown embraced the idea
as a keynote in his program for change and created an Office of
Appropriate Technology to promote "AT" in the state. During
the presidency of Jimmy Carter a number of agencies of the federal government paid at least some attention to the idea. For a
time, Carter's antipoverty agency, Action, featured "appropriate
technology" and "self-sufficiency" slogans in its programs to
help the poor. At the urging of Senator Michael Mansfield, federal funds were allotted to found a National Center for Appropriate Technology located in Butte, Montana.

Little noticed in all the ferment was a basic conceptual problem. Any notion of "appropriate" technology is meaningless
until one answers the question, Appropriate to what? As applied

to the Third World, the answer was clear enough; judgments about appropriateness would come from specific cultural and environmental settings. Agricultural production techniques well suited to one country might not be desirable in another; each society would have to determine what means are appropriate to its needs. As applied to Western industrial societies, however, the idea had a much different significance. After all, these were societies whose dominant cultural norms had led them astray. If "appropriate technology" was to have any significance at all, it would have to challenge these norms and suggest new ones. But where would this new understanding be found? And how would it persuade those already committed to orthodox forms of technical and economic practice?

Political and Intellectual Origins

APPROPRIATE TECHNOLOGY does not have a long, coherent tradition in the same way, for example, that contemporary labor movements find their origins in working class struggles of the past. Modern history contains recurring, but widely scattered attempts to find social and technological alternatives to the dominant patterns of industrialism. Ideas and experiments in this vein can be found in the writings of Robert Owen and other nineteenth-century utopians, William Morris, Peter Kropotkin, Gandhi, the Spanish anarchists, the British Guild Socialists, the cooperative movement, Lewis Mumford and the American decentralists, the followers of Rudolf Steiner's biodynamic agriculture, back-to-the-land movements, and whole generations of tinkers and crackpot inventors.[4] While movements of this kind have sprung up repeatedly during the past two centuries, they have not, by and large, been able to sustain their work or to leave a legacy of accomplishments for the next generation of reformers.

The more immediate sources of interest in a radical approach to technology, however, grew from social movements of the 1960s and early 1970s. Many of the same concerns and passions that fueled activism in civil rights, New Left politics, anti-war protests, counterculture, and environmentalism led eventually to a critical reexamination of the foundations of modern industrial society. This was territory that few persons involved in the early stages of 1960s radicalism in the United States were eager to enter. Civil rights and Vietnam War protests in 1962 through

1967 seldom identified either capitalism or technology as the cause of the problems they sought to address. While the workings of the "machine" or "system" were frequently mentioned as deeper maladies underlying social injustice and international aggression, such terms were employed metaphorically and did not point to anything specifically technological. A readiness to focus upon machines, techniques, chemicals, experts, and large-scale sociotechnical systems as sources of evil arose later on as social conflict in the United States and other industrialized nations intensified in the late 1960s. Searching for a common element in the maladies that angered and vexed them, activists began to focus upon modern technology as an increasingly obvious and perhaps too convenient key. Campus demonstrations against U.S. chemical corporations vented students' rage against the napalm dropped on helpless mothers and children in Vietnam. The lethal hardware of the modern battlefield, presented with shocking clarity each night on network television newscasts, called attention to the war's technical side. Those who could not bring themselves to study the roots of the Vietnam War in the workings of American foreign policy or the goals of international capitalism were able to focus upon the instruments of destruction and upon the persons who manned and directed the B-52s and helicopter gunboats. Soon the link between destructive herbicides in Southeast Asia and the herbicides, pesticides, and industrial pollutants so recklessly dispensed at home gave new meaning to the word "environment." For fully a century and a half prior to the Vietnam War, the advance of technology and the horizons of human progress had seemed one and the same. Now it appeared that the most sophisticated products and best practitioners of scientific technology had somehow been enlisted in a war against progress—toward the annihilation of everything decent, humane, and just.

This obsession with technology arose at about the same time, 1969 and after, that a general disillusionment with politics had begun to erode the energy and commitment of those active in protest movements. Police and army shoot-ups of marches and political headquarters, the obvious presence of agents provocateur, and the arrest and the trial of activists for alleged crimes led to a general cooling of the New Left fervor.[5] The weariness that many had come to feel in the endless rounds of meetings, rallies, and unsuccessful election campaigns had also taken its toll. It was during this period that many in the United States dropped out of political activity and began a certain kind of sociotechnical tinkering: roof gardens, solar collectors, and windmills became a focus of community action. Politics that had become too dangerous or depressing when faced with a showdown with real power tried to reestablish itself as a quest to reform "the system" through social and technical invention.[6]

Evidence of this transition is preserved in the contents of two catalogues of American radicalism published late in the decade: *The Movement Toward A New America*, assembled by Mitchell Goodman, and *The Whole Earth Catalog*, edited by Stewart Brand and his colleagues.[7] Goodman's volume gathered together hundreds of handbills, pamphlets, articles, and news stories that reflect the political activities of the 1960s. Rent strikes, The Black Panthers, Students for a Democratic Society, anti-war demonstrations, activities of radical feminists, and the like are described in the actors' own words and in the writing of sympathetic observers. He portrayed a spirited, many-faceted, nationwide movement challenging the oppressive hold of an established order. Published in 1970, the book was actually a last gasp. By the time it became possible to arrange a snapshot of all the different themes and actors, the movement had already fallen into decline. *The Movement Toward a New America* offered itself as a handbook for people just getting under way. We can now appreciate it as a scrapbook for those who were just getting out.

Although actually published before the Goodman collection, *The Whole Earth Catalog* expressed a tendency then gaining momentum rather than withering away. Its vision was that of a groovy spiritual and material culture in which one's state of being was to be expressed in higher states of consciousness and well-selected tools. Citizenship had at that time, 1968 to 1969, become an onerous burden for some. The Weathermen denounced the United States of America, declaring that they were actually soldiers in the revolutionary armed forces of the Third World.[8] In contrast, Stewart Brand's book consoled its readers with the idea that they were citizens of the planet earth and its global systems: hippie environmentalist spacemen in the tradition of Buckminster Fuller. Here the obsession with technology, especially *good* technology, a theme almost totally absent from Goodman's compendium, made a strong appearance. In its statement of purpose the *Catalog* announced that "a realm of intimate, personal power is developing—power of the individual to conduct his own education, find his own inspiration, shape his own environment, and share his adventure with whoever is interested." It explains that an item is listed "if it is deemed: (1) useful as a tool; (2) relevant to independent education; (3) high quality or low cost; (4) easily available by mail."[9]

If you were moving back to the country and needed a good sturdy cider press, *The Whole Earth Catalog* would tell you where to purchase one. Whole categories of small (and sometimes large) producers were brought in touch with a new generation whose life-style and habits of consumption had begun to change. *The Whole Earth Catalog* assumed that throngs of people would be moving off into small, humanly nurturing, economically self-sustaining communities that fit into a new complex world system destined to save the earth from the destruction of overindustrialization. In this vision choices about the right technologies—both useful old gadgets and ingenious new tools—mattered greatly; choices about politics mattered little. Preferring brief, enthusiastic descriptions of items for sale, *The Whole Earth Catalog* avoided essays on controversial topics of social, political, or even environmental concern. The catalogue-browsing consciousness of the New Age was not one that wanted to be bothered by well-reasoned arguments.

The growing fascination with technology among those involved in movements of the late 1960s was nourished by the writings of well-known European and American intellectuals. Starting from very different points of view, Lewis Mumford, Paul Goodman, Herbert Marcuse, Theodore Roszak, and Jacques Ellul carried the subject of modern technological society and the technical mentality to the foreground of social criticism.[10] Marcuse's *One-Dimensional Man* (1964) portrayed both capitalist and socialist societies as components of a vast, repressive technological civilization that was bringing every aspect of humanity under its control. Mumford's *The Myth of the Machine: The Pentagon of Power* (1970), pessimistic climax of a lifetime of commentary on material culture, judged the promise of modern technics betrayed by the destructiveness of authoritarian megatechnics and the spiritual hollowness of expertise. Ellul's *The Technological Society* (1964) provided its readers with an extreme statement in the same vein, arguing that every aspect of twentieth-century life—economics, politics, symbolic culture, individual psychology, and so forth—had fallen under the domination of *la technique*. Such books were read and widely talked about by those who found more orthodox modes of social analysis inadequate to express the troubles they saw in the modern world. Either there was something abominable in modern artifice itself, a position strongly argued by Jacques Ellul, or the particular instruments most commonly used in the modern age were simply the wrong ones, wrong in the sense that they generated destruction so vast as to undermine the very benefits of technological productivity. It was the latter belief, however reasonable or unreasonable it may be, that eventually helped spawn the idea that an alternative or appropriate technology was something that one could expect to discover or invent.

An uneasiness with conditions of modernity has been a consistent theme in Western literature since the beginning of the industrial revolution.[11] Thomas Carlyle's "Signs of the Times" (1829), for example, already contains many notions central to mid–twentieth-century criticisms of technological society: awe at man's domination of nature, unease at the disruption of tradition, disgust with the regimentation of workers, rage at the injustices of the industrial economy, and anxiety at the loss of a moral center in the face of technical advance. "Men are grown mechanical in head and heart, as well as in hand," Carlyle exclaims. "They have lost faith in individual endeavour, and in natural force, of any kind. Not for internal perfection, but for external combinations and arrangements, for institutions, constitutions—for Mechanism of one sort or other, do they hope and struggle. Their whole efforts, attachments, opinions, turn on mechanism, and are of a mechanical character."[12]

The tendency to obliterate distinctions between technology and other social phenomena, a conceptual expedient characteristic of Ellul, Marcuse, and Roszak, is fully present in Carlyle's version of the story. In search of a comprehensive account of the malady and a diagnosis of its cause, critics of industrial society typically arrive at conclusions that deny any chance of practical remedy. Specific ills in industrial civilization are said to be rooted very deeply in human aggressiveness, the machine mentality, the essence of the subject/object split, the obsession of the West with *la technique*, rational thought, the workings of the second law of thermodynamics, or in some similarly intractable source. Thus, over many generations critics of industrialism and modern techniques have been unable to propose cures for the host of difficulties they so vividly describe. Their lament is clear enough. But the eloquence of criticism—and perhaps this is a property of criticism—is matched by a poverty of practice.

Since technology is the sphere in which "it works" is paramount, an uneasiness with problem solving seems to undermine the very basis upon which nontechnical thinkers have credentials to talk about technical matters at all.

The shortcomings of the critics are obvious. But it is also true that their concerns are among those most urgent in nineteenth- and twentieth-century social thought. Hopes that an "appropriate technology" might be found arose at a time in which these concerns had been renewed, restated for a new generation, and popularized for a mass audience. Surprisingly enough, during the late 1960s and early 1970s the sense of futility and despair so often characteristic of speculative writings on technology was suddenly replaced by an extravagant optimism. Philosophers and political theorists began to ponder the issue of what the technology of an emancipated society should look like. They began to wonder openly, Who might take up the tasks of realizing such a vision? Upon what occasion might their opportunity come?

Herbert Marcuse's *An Essay on Liberation* (1969), for example, offers an unabashedly utopian response to the dismal situations sketched in *One-Dimensional Man* (1964). He writes, "Freedom indeed depends largely on technical progress, on the advancement of science. But this fact easily obscures the essential precondition: in order to become vehicles of freedom, science and technology would have to change their present direction and goals; they would have to be reconstructed in accord with a new sensibility—the demands of the life instinct. Then one could speak of a technology of liberation, product of a scientific imagination free to project and design the forms of a human universe without exploitation and toil."[13] Marcuse argues that such steps would be possible "only after the historical break in the continuum of domination" and tries to spell out the conditions of moral decay and economic and political disorder that could undermine and eventually topple the repressive society. Impetus for the change will come from a combination of traditional working class grievances, middle-class rebellion at conditions of surplus repression, and the quest for freedom and social justice among Third World peoples and the minorities of industrialized nations. The essay envisions "not the arrest or reduction of technical progress, but the elimination of those features which perpetuate man's subjection of the apparatus and the intensification of the struggle for existence."[14] Pointing to a social reconstruction that includes political judgments about technological design features, Marcuse had begun building a bridge between the central themes of Frankfurt School critical theory and the possibility of an alternative technology.

In a move intellectually comparable to Marcuse's, Theodore Roszak's *Where the Wasteland Ends* (1972) sets out to propose an alternative society, the "visionary commonwealth," that could overcome the destructive tendencies of technocratic civilization. Roszak's previous work, *The Making of the Counterculture* (1969), focused upon the pervasive sickness of modern militarism, urbanism, consumerism, bureaucracy, and the technocratic mentality, but ended with a song of praise to shamanism and the call to transcend the subject/object split.[15] It was not much of a program for action. The later book, however, answers its pleas for spiritual enlightenment with speculative proposals for decentralization, de-urbanization, the creation of economically self-sufficient communities, and innovation toward a broad range of alternative technical and social forms. Calling attention to the history of theory and practical experiments in communitarianism and anarchist socialism, Roszak imagines "the proper mix of handicraft labor, intermediate technologies, and necessarily heavy industry . . . the revitalization of work as a self-determining, non-exploitative activity . . . a new economics elaborated out of kinship, friendship and co-operation . . . the regionalization and grass roots control of transport and communication . . . non-bureaucratized, user-developed, user-administered services."[16] He admits that in a world such as ours these lovely plans are likely to sound foolish. "I can think," he explains, "of forty reasons why none of [these] projects can possibly succeed and forty different tones of wry cynicism in which to express my well-documented doubts. But I also know that it is more humanly beautiful to risk failure seeking for the hidden springs than to resign to the futurelessness of the wasteland."[17]

In retrospect it may seem that these writers were naive, that they underestimated the power of the dominant institutions of the late twentieth century. But, clearly, that is not true. Roszak, Marcuse, Mumford, and others writing along similar lines were perfectly aware of the Pentagon, CIA, transnational business firms, and other megalithic high-technology organizations. It was not naivete their writings expressed, but rather total contempt for these institutions combined with a sense of powerlessness. To avoid the cynicism and gloom toward which their thinking carried them, it was necessary to perform a high-wire act along very slender threads of hope.

The appropriate technology movement in industrialized countries set out to walk the same tightrope. Stemming from the decline of radical politics and from an obvious next step in the critique of technological society, its true purpose was not to produce energy from renewable resources, but to generate the hope of social renewal from the winds of despair.

The Hard and The Soft

BUT COULD the study of alternative technologies lead to a fruitful critique of modern society? Why not an alternative economics, for example? Indeed, in Europe and the United States during the late 1960s and 1970s, many young, politically oriented intellectuals turned to Marxian economics as a way to deepen their understanding of social power. The founding of the Union for Radical Political Economics in the United States came at about the same time as the start of The New Alchemy Institute and reflected a similar sense of expansive intellectual horizons. Disciplinary orthodoxies, so long silent on many of the issues that seemed most urgent, were rejected in favor of a fresh approach to the study of material culture. Thus, the radical political economists opened fire on the neo-Keynesian philosophy taught in universities because it accepted the existing relations of production in capitalism. Any science of economics worthy of the name, they argued, would have to address basic questions about social structure and the just distribution of wealth.[18]

Similarly, the realm of "technology" was suddenly, surprisingly thrown open to imaginative research and theorizing. The ethical, ecological, political, and even metaphysical dimensions of the technical disciplines suddenly became topics of widespread concern. This was no longer solely the work of intrepid outsiders without suitable credentials. As an interest in appropriate technology spread during the mid-1970s, a number of insiders, inspired technical specialists, began to investigate novel styles of interpretation in science and engineering. As the rosters of such organizations as the Brace Institute, Farallones Institute, Intermediate Technology Development Group, and State of California Office of Appropriate Technology made clear, a good number of those working to open these boundaries had excellent backgrounds in biology, physics, electrical engineering, architecture, and other scientific and technical fields. They presumably knew "the hard stuff" and what the dominant orthodoxy has to say about how one properly follows one's profession in the "real world." But recognizing a need to set their knowledge and skills to fundamentally new purposes—purposes outside those normally recognized by the corporations, foundations, government, and universities—many scientifically trained activitists tried to redefine the context in which scientific research and technological innovation could take place.

The period's most colorful record of the attempt to do basic and applied science in a new vein is contained in the *Journal of the New Alchemists*. Its volumes include reports on aquacultural and agricultural experiments, diagrams of solar and wind devices, tables of data on current research, and other features that reflect the canons of modern scientific methodology. Interspersed with these, however, are poems, philosophical essays, drawings, and highly personal accounts of the experience of the institute's members. Introducing the article by aquaculturalist Bill McLarney, "Studies of the Ecology of the Characid Fish *Brycon Guatemalensis* in the Rio Tirimbina, Heredia Province, Costa Rica with Special Reference to its Suitability for Culture as a Food Fish," the editor comments on McLarney's research procedures: "Clad in sun hat, fisherman's vest, shorts, and running shoes he pursued his quarry through the water with Chaplinesque élan. In both his science and his writing his standards of excellence are irreproachable but being around the work actually in process is to be part of an on-going comedy."[19]

Presenting their work in this manner, the New Alchemists risked their credibility within the community of scientists and engineers. The accepted form of "objectivity" in scientific and technical reports (one can also include books and articles in social science) requires that the prose read as if there were no person in the room when the writing took place. While the New Alchemists recognized the standards of experimentation and evidence prevalent in modern research, they rejected notions of objectivity that conceal the social purposes of science. Their aims were explicitly stated on the inside cover of their journal. "We seek solutions that can be used by individuals or small groups who are trying to create a greener, kinder world. It is our belief that ecological and social transformations must take place at the lowest functional levels of society if people are to direct their course toward a saner tomorrow."[20] According to a story told by The New Alchemy Institute cofounder John Todd, the institute formed when a group of young biologists finally grew tired of attending prestigious international conferences at which calamities for the planet through pollution, overpopulation, and other environmental dangers were blithely predicted by experts whose research commitments seemed to preclude any attention to how such disasters might be averted.

The rise of a growing number of research institutes, small firms, government agencies, universities, philanthropic organizations, and independent individuals claiming to do appropriate technology was not a sign that such people shared a common philosophy. In Amory Lovins' view of the matter, for example, appropriate technology was fully compatible with capitalism, albeit a capitalist system reorganized to provide sufficient investment funds for the "soft energy path."[21] David Dickson, British writer on science and technology, saw the possibilities from a Marxist perspective, arguing that "alternative technology" represented a new chapter in the history of socialism.[22] Murray Bookchin, veteran political activist of the American left and onetime teacher of appropriate technology of Goddard College, joined others, including Theodore Roszak and Ivan Illich, in an interpretation that claimed the movement as a rebirth of communitarian anarchism.[23] Not to be outdone, the National Academy of Sciences commissioned economist Richard Eckaus to study the topic within the dry and predictable criteria of 1950s social science: economic growth, transfer of technology, maximization of output, political development, and the like.[24] Indeed, as people from different backgrounds and persuasions sought to move from staid orthodoxies toward new understandings, there were numerous attempts to attach the insignia of one's own club. It soon became clear that what was happening was less the oft proclaimed "paradigm shift" than a grinding of ideological gears.

In an honest desire to bring order out of conceptual chaos, a number of proponents argued that "appropriate" or "alternative" or "soft" technology was a uniform program with a logically consistent set of characteristics. They offered long lists of criteria that were supposed to describe a coherent ideal. One such list, prepared by Robin Clarke of Biotechnic Research and Development in England, gave thirty-five criteria for "soft technology." It was said to "form a coherent system that can only be interpreted as a whole, and loses much of its sense when reduced to fragmented components." Thus, "soft technologies" were characterized as "ecologically sound, small energy input, low or no pollution rate, reversible materials and energy sources only, functional for all time, craft industry, low specialization, . . . integration with nature, democratic politics, technical boundaries set by nature, local bartering, compatible with local culture, safeguards against misuse, dependent on well-being of other species, innovation regulated by need, steady state economy, labour intensive, . . . decentralist, general efficiency increases with smallness, operating modes understandable to all," and so forth. In the other list characteristics of "hard technology" were described as "ecologically unsound, large energy input, high pollution rate, non-reversible use of materials and energy sources, functional for limited time only, mass production, high specialization, . . . alienation from nature, consensus politics, technical boundaries set by wealth, world-wide trade, destructive of local culture, technology liable to misuse, highly destructive to other species, innovation regulated by profit and war, growth-oriented economy, capital intensive, . . . centralist, general efficiency increases with size, operating modes too complicated for general comprehension," and the like.[25]

Inevitably, Clarke's typology and all similar ones were bound to fail. Nothing in Western philosophy—or in all of human experience for that matter—suggests that we can arrange the good and the bad in simple lists. If one pursues an ideal of justice far enough, for example, it may well begin to conflict with one's own best notion of freedom. By affirming policies compatible with local culture, there is no guarantee that one will promote more democratic politics. Neither the intricacies of theory nor the evidence of historical practice gives us any reason to believe that the ideals of self-sufficiency, community, safety, diversity, efficiency, and the like can be easily gathered under one umbrella. It is not merely that such wishes are not feasible in practice, that certain facts in this world stand as barriers to their realization. More important, the set of criteria upon which this vision of good technology rests is filled with conditions that may not be compatible. Hence, it is not obvious that decentralized technologies are necessarily environmentally sound; the widespread use of wood-burning stoves has produced very high levels of pollution in some localities. It is not clear that labor-intensive systems provide "work undertaken primarily for satisfaction"; some of history's worst sweatshops were operations of that kind. Anyone who has lived in a small town knows that small communities may be the last to consider "diverse solutions to technical and social problems." If one had to select a theme song for appropriate technology, it would be wise to avoid rousing victory marches and settle for Charles Ives' "The Unanswered Question."

The New Age Begins

DESPITE THE FACT they contain contradictory ideals, lists such as Robin Clarke's do try to answer the question I noted at the beginning: How can one define "appropriate" technology for an advanced industrial society? During the 1970s definitions of this sort acknowledged many issues of widespread public concern—"limits to growth," the shortage of fossil fuels, worries about how to solve world population and hunger problems, a growing alienation from government, and suspicion of large public and private bureaucracies. But there was always something strangely incomplete in this collection of issues. The astonishing fact never accounted for was why so many members of the North American and European middle class, arguably those best served by modern technological society, should have become fascinated with "appropriate" technology at all.

Among the places where the symbolism and underlying motives of the movement were most clearly visible were the fairs, expositions, and festivals that heralded the coming of the so-called New Age. In much the same way that the great world's fairs of the nineteenth and twentieth centuries offered an opportunity to witness our civilization's grandest (and often least realistic) hopes for science and industrial technology, New Age expositions of the sort held in Los Angeles, San Francisco, Vancouver, Boston, and other North American cities revealed a great deal about the deeper motives of appropriate technology proponents.

One such fair that I was able to visit and study in person was the New Earth Exposition held in Boston in early May 1978. The theme of the exposition was "Living Lightly on the Earth." Its advertising promised it would be "a showcase for environmentally creative individuals and business demonstrating viable alternatives in the field of: Energy, Personal Growth, Food, Transportation, Shelters, Gardening, Recycling, Wilderness Skills, Ecology, and Publishing." On the floor of the huge

74

Commonwealth Pier Exhibition Hall were hundreds of small (and some not so small) organizations showing their wares and passing out literature. I was astonished at the sheer number of groups represented and surprised at how far they had come in developing their goods and services. In the category of wood-burning stoves alone there were no fewer than a dozen entrants. Choosing a wood-burning stove had become as difficult as selecting a new car.

In the general run of items and ideas displayed at the expo, those which most clearly fit the category "appropriate technology"—solar collectors, owner-built homes, energy-saving devices, windmills, wood stoves, solar greenhouses, and other kinds of new-fangled and old-fashioned hardware—were, oddly enough, also those which most readily fit the model of ordinary American respectability. Along with their explanations about the environmental benefits to be gained from using these devices, the exhibitors stressed the virtues of sound engineering, thrift, and good business sense. At one and the same time the salesmen appealed to very conservative as well as radical perceptions of what an intelligent person ought to be doing.

Other parts of the show catered to the visitors' tendencies toward escapist consumerism and spiritual self-indulgence. A large cluster of exhibits advertised new toys for the young, professional middle class: jogging shoes, racing bicycles, hot tubs, backpacks, tents, hang gliders, and herbal cosmetics. The traditional American notion that freedom means something like being able "to hit the road whenever I feel like it" was fully represented, with hiking equipment and sleeping bags replacing the automobile as vehicles for realizing this fantasy.

Promoters of alternative health care measures were also conspicuously present. An amazingly large number of booths presented the wonders of New Age medicine—acupuncture, Shiatsu massage, macro vitamin therapy, herbal remedies, iridology, and other such practices—a new industry to compete with the medical establishment. Representing an adjacent field of concern, advocates of various spiritual disciplines sought converts from among passersby—yoga, transcendental meditation, Arica, Sufism, T'ai Chi Chuan, pyramid power, and several kinds of spiritual communities. The market was open for anyone buying. Explicitly political causes, however, were pretty much absent from the event. Environmental activists and organizations opposing nuclear energy or trying to save the whales were just about the only groups that bothered to secure the easily obtainable exhibitors' invitations. Evidently, the New Earth in its New Age did not recognize a need for political organization.

Beyond the honest concern to preserve the natural environment, beyond even the wish to rediscover community, personal autonomy, or both, there was an obvious yearning contained in the fair's symbolism. Underlying the vast majority of exhibits and not too far below the surface of their claims was the simple need to overcome tension, the stresses and strains of modern civilization. People thoroughly enmeshed in the demanding business of keeping our technological society running were beginning to hedge their bets. Through bicycling, transcendental meditation, running, organic food, massage, backpacking, yoga, and a host of ingenious means they would seek the relaxation and peace of mind that everyday life cannot provide. Yes, technologies of a certain kind would be required. But they were not inventions such as solar hot-water heaters. Wanted instead were devices and techniques that might alleviate the pressures that normally befall professionals toiling in banks, insurance companies, and bureaucracies. Rather than attempt to change the structures that vexed them, young Americans growing older were settling for exquisite palliatives. If the 1960s proclaimed, "Let's see if we can change this society," the 1970s answered, "Let's get out of this skyscraper and go jogging!"

Contained in the frantic quest for physical vitality and self-realization was a growing distrust of the goods and services available in our society as well as a distrust of the professionals and organizations that provide them. Who needs doctors? Do your own health care. Who needs architects and contractors? Build your own home. Who needs the utilities? Generate your own energy. The desire for "self-sufficiency," long regarded as a virtue in Western culture, had been clouded by a profound resentment. Cramped by inflation, feeling pressured in unsatisfying jobs, fearful of cancer and heart attacks, bored with the products of the consumer economy, and unwilling to escape to the once glorious idyll of suburban living, a substantial number of those who had "made it" in modern society were, for the moment at least, restless.

Demonstration Models

THE IDEALS of appropriate technology and appetites of New

Age consumerism point to a crucial flaw in the dreams of modern materialism. Although society achieves unparalleled wealth, an unmistakable corruption begins to afflict the very objects and institutions of industrial culture. Because the automobiles, appliances, energy systems, prepared foods, cosmetics, and other products and services have been engineered to corrupt specifications, the good life begins to look like a colossal pile of junk. The promise of an artificial paradise in which all human needs will be satisfied gradually dissolves.

In this sense an important mission of the appropriate technologists was to renew the war against qualities that William Morris had long ago named "the shoddy." If a way could be found to reform the obvious inadequacies of material things, it would amount to a reform of society itself. This was, in fact, often proposed as a solution to the problem noted earlier: How would appropriate technology persuade those already committed to orthodox forms of technical and economic practice? The answer: give them a superior product.

One of the clearest statements of this point of view was offered in an insightful and amusing book, *How Things Don't Work*, by industrial designers Victor Papanek and James Hennessey. With a delight that others reserve for watching old Marx Brothers movies, they report dozens of examples of deplorable screwups in the design and engineering of everyday tools and consumer goods. "Between 1966 and February 1975 the automobile industry recalled 45,700,000 automobiles for inspection or repair," they observe to indicate how bad things have gotten.[26] Beneath a photograph of two commercial cheese graters they note, "On the left, the inexpensive, efficient, and nearly indestructible one which will work left- or right-handed." On the right, the "improved model, which is right-handed only and, after some months of use, grinds its own plastic coating into the food."[27]

Citing a wide range of similarly appalling cases, the book locates some of the causes of shoddy industrial design and manufacture. Unwise cost cutting in the name of profit ruins the quality of many goods. Alienated assembly line workers take secret revenge upon the automobiles and other products that pass by them on conveyor belts. The effects of advertising and showcase packaging decrease the demand for previously useful items. Senseless so-called improvements justified by norms of "convenience," "ease of handling," and "customer comfort" drive up the cost but degrade the utility of home tools and appliances. As one reads Papanek and Hennessey's observations, it begins to seem they have successfully taken the criticisms of Mumford, Ellul, Marcuse, and Illich about technological society and transformed them into a set of perverse design criteria for modern invention. They seem only slightly less bizarre when we notice that they include many of the criteria upon which our industrial economy is actually founded.

Papanek and Hennessey's response to these maladies reveals not only their own sense of what is to be done, but also reflects a model of social change implicit in the writings and projects of appropriate technologists of all persuasions. "Massive problems faced by workers and users need innovative remedies," they explain, and offer a brief survey of frequently proposed nostrums. "There is the capitalist approach (make it bigger), the technocratic one (make it better), the 'revolutionary' solution (portray the problem as an example of an exploitative system) and the pre-industrial romantic fallacy (don't use it; maybe it will go away by itself). We propose a fifth alternative response: *Let's invent a different answer.*"[28]

Donning the mantle of an old-fashioned tinker/designer/ crackpot inventor, Papanek and Hennessey describe an array of clever techniques and devices that they and their colleagues around the world have created. A pedal-driven lawn mower, fold-up bathtub, fold-up bicycle, spillproof scooter, do-it-yourself emergency service vehicle, and other contraptions "fitted to human scale" are depicted in drawings and photographs. Innovations in hardware, they point out, must be matched by innovations in ownership and use. "Few tools in our society are designed for communal (or shared) ownership. If they were designed for sharing, rather than for individual use, we believe they would change structurally, mechanically and in material composition. When reel-type push mowers were still in use, for example, they were more frequently borrowed or shared than today's electric or gasoline-driven mowers for the simple reason that they were robust, uncomplicated and difficult to break."[29]

Papanek and Hennessey's guiding maxim—"Let's invent a different answer"—echoes a familiar apothegm usually attributed to Ralph Waldo Emerson (although it is not certain that he said exactly these words): "If a man can write a better book, preach a better sermon, or make a better mousetrap than his neighbor, though he builds his house in the woods the world will make a path to his door."[30] This is, of course, the traditional American notion about how inventions change the world. A good idea or improved tool is bound to catch on. In fact, as Emerson evidently wished to say, if one's scheme is attractive enough, there will be no stopping it. Ingenuity creates its own demand.

As scholars Arthur Bestor and Dolores Hayden have observed, nineteenth-century American utopians held much the same conviction about how their experiments might eventually transform society. The utopians believed their technical inventions and social innovations would have a strong appeal to an age undergoing rapid change. Communities such as those at New Harmony and Oneida saw themselves perfecting what Bestor calls "patent office models" of the good life.[31] In the same way that ordinary people would eagerly accept new improvements in farm machinery if a convincing demonstration were given them, so would they be willing to embrace the principles and devices of utopia if a successful working model could be built and maintained somewhere in the world.

Insofar as they had a coherent idea of how their labors would change the world, the appropriate technologists usually entertained the better mousetrap theory. A person would build a solar house or put up a windmill, not only because he or she found it personally agreeable, but because the thing was to serve as a beacon to the world, a demonstration model to inspire emulation. If enough folks built for renewable energy, so it was assumed, there would be no need for the nation to construct a system of nuclear power plants. People would, in effect, vote on the shape of the future through their consumer/builder choices. This notion of social change provided the underlying rationale for the amazing emphasis on do-it-yourself manuals, catalogues, demonstration sites, information sharing, and "networking" that characterized appropriate technology during its heyday. Once people discovered what was available to them, they would send away for the blueprints and build the better mousetrap themselves. As successful grass-roots efforts spread, those involved in similar projects were expected to stay in touch with each other and begin forming little communities, slowly reshaping society through a growing aggregation of small-scale social and technical transformations. Radical social change would catch on like disposable diapers, Cuisinarts, or some other popular consumer item.[32]

79

The inadequacies of such ideas are obvious. Appropriate technologists were unwilling to face squarely the facts of organized social and political power. Fascinated by dreams of a spontaneous, grass-roots revolution, they avoided any deep-seeking analysis of the institutions that control the direction of technological and economic development. In this happy self-confidence they did not bother to devise strategies that might have helped them overcome obvious sources of resistance. The same judgment that Marx and Engels passed on the utopians of the nineteenth century apply just as well to the appropriate technologists of the 1970s: they were lovely visionaries, naive about the forces that confronted them.[33]

Far and away the most grievous weakness in their vision, however, was the lack of any serious attention to the history of modern technology. Presumably, if the idea of appropriate technology makes sense, one ought to be able to discover points at which developments in a given field took an unfortunate turn, points at which the choices produced an undesirable instrumental regime. One could, for example, survey the range of discoveries, inventions, industries, and large-scale systems that have arisen during the past century and notice which paths in modern technology have been selected. One might then attempt to answer such questions as, Why did developments proceed as they did? Were there any real alternatives? Why weren't those alternatives selected at the time? How could any such alternatives be reclaimed now? In some of their investigations in agriculture and energy, appropriate technologists began to ask such questions. But by and large most of those active in the field were willing to proceed as if history and existing institutional technical realities simply did not matter. That proved to be a serious shortcoming. It meant that many of their projects were irrelevant to the technical practices they hoped to challenge.

The New Age Ends

THE DEMISE of appropriate technology as a living social movement can be dated almost to the hour and minute. It occurred early in the evening of Tuesday, November 4, 1980, when it became clear that Ronald Reagan had been elected President of the United States. That event signaled the end of a favorable climate of opinion for public discussions of energy policy, energy conservation, and alternative energy technologies that had characterized the presidency of Jimmy Carter. "Let the market work," was now to be the central premise of government policy. During the era of Reaganomics, most planning of nonmilitary technologies moved back to the private sector, back into the hands of business corporations, and away from public scrutiny.

In itself, the coming of a new presidential administration should not have spelled the doom of appropriate technology. Although some of the movement's research and development projects had been funded by government agencies, its basic thrust did not depend solely upon bureaucratic sponsorship. But in ways the appropriate technologists little understood, the reception given their ideas, including their most radical ideas, was a function of national politics and its shifting fortunes. As long as there were liberals or moderates in key posts in the Department of Energy, Environmental Protection Agency, and other government offices, appropriate technologists saw at least some chance of finding an audience, winning influence, and promoting their better mousetraps. With that audience gone, ideas for solar collectors, windmills, aquaculture ponds, community-owned factories, and the like became orphans overnight. The early 1980s might have been a time to regroup and rededicate. Instead it became a time of retreat. Within a period of months after Ronald Reagan took office, appropriate technology disappeared as a concept featured in conferences, media reports, government surveys, academic programs, and paperback books. It was an idea whose time had come . . . and gone.

Indeed, the New Age itself, which many expected to endure for centuries, had lasted a scant four years. This sad fact was duly noted by the nation's leading countercultural historian, antinuclear activist Harvey Wasserman. In his book *America Born and Reborn*, Wasserman develops a cyclical theory of history to explain the pattern of national events. One of his cycles, for example, begins with the election of Andrew Jackson, continues with the Transcendental Revolt, moves on to the Civil War, and culminates in the Gilded Age. An acceleration of events in modern history has, according to Wasserman, caused these cycles to become shorter and shorter. Thus, the cycle that begins with Jimmy Carter's proclamation of a "New Spirit" develops into the "Solar Awakening," runs into trouble with the Iranian hostage crisis, and finally reaches its denouement during the presidency of Ronald Reagan. The book offers hope, however, forecasting yet another era about to begin, "The Greatest Awakening," an enduring time of spiritual renewal in which war and injustice will be overcome, ushering in "our next evolutionary leap— into the Computer Age, and into the communal frontiers of the mind and spirit that await our full human potential."[34] Heaven knows, we need a new epoch. But having watched the mercurial rise and fall of the last New Age, one should be careful not to step out for popcorn.

In some form or other, appropriate technology will likely endure. The New Alchemists and Intermediate Technology Development Group are still at work trying to develop tools that are locally useful, environmentally sound, and compatible with the most noble social ideals. Here and there around the world one finds research institutes, nonprofit organizations, and university programs that carry on the tradition. The National Center for Appropriate Technology in Butte, Montana, hampered by shifting political priorities and funding cuts, continues its work but with few demonstrable accomplishments. Unfortunately, California's Office of Appropriate Technology was abolished as soon as a conservative Republican governor took office. The strongest continuing effort to translate the movement's ideals into a comprehensive program for social change is put forth by "the Greens," the progressive, ecologically oriented political parties that sprung up in Europe in the early 1980s.[35]

The legacy of appropriate technology cannot be found in the catalogues of inventions and techniques it left behind. If the truth be told, few of them worked very well in practice. "Our windmill's being repaired today and the solar collector should be operational next week" was a standard explanation visitors heard at the demonstration sites. Seldom was it possible for any of the research groups to move through the various stages of development, debugging, and deployment that any new technology needs. Too often devices were announced as ready and working that actually needed much more time for testing. Those who believed that a new brand name had come on the market— "The next time you go to the store, be sure to ask for *appropriate* technology."—were bound to be disappointed.

The accomplishments of appropriate technologists, ones that will be interesting to study in future decades, are located in the realm of ideas. For better or worse, they did challenge many of the key premises in the modern technical orthodoxy. They helped broaden the meaning of such categories as "efficiency," "rationality," "productivity," "cost," and "benefit" and added fresh (if not altogether novel) criteria of judgment—"human scale," "the interconnectedness of things," "second law efficiencies," "sustainability," and the like—to the range of considerations that engineers, technicians, agriculturalists, planners, and consumers ought to take seriously in making choices. To a certain extent these notions have now become common currency. It is more likely now that decision makers and the public will pause to think about such matters as the ecological soundness of a proposed practice before barging forth. And it is somewhat less likely that some destructively narrow but commonly used categories of technical and economic analysis will continue to cripple our powers of judgment. In that sense the movement succeeded admirably.

Appropriate technology arose in a world of bizarre inversions that are still the source of our civilization's perplexity. While millions of people around the globe go without adequate food, shelter, or work, hundreds of billions are spent each year on increasingly sophisticated, increasingly lethal armaments that are said to "secure the peace." Irreplaceable resources needed by future generations are mined and quickly used up in the belief that somehow "the market" will produce an inexhaustible supply. Nations ignore or revise downward their standards of environmental quality because pollution controls are thought to be incompatible with a "sound economy." In situations like these we become used to accepting deplorable conditions as normal. To be realistic, to get things done, and to get on with one's career almost require that a person become an enemy of a free humanity and a healthy biosphere.

The condition we face is much like that described in Bertolt Brecht's little play, *The Exception and the Rule*. On Brecht's stage a handful of characters wander through a pattern of actions that show a moral universe turned on its head. What is good is made to appear evil; justice and injustice trade places. A coolie attempts to do a good deed. He is killed by his employer who sees the coolie's gesture as a threat from a class enemy. The murderer is placed on trial but acquitted in a judgment that finds his behavior perfectly reasonable under the circumstances. Brecht is never one to let his audience miss the point. At the beginning of the drama his actors exclaim[36]:

> Inquire if a thing be necessary
> Especially if it is common
> We particularly ask you—

Not on that account to find it natural
Let nothing be called natural
In an age of bloody confusion
Ordered disorder, planned caprice,
And dehumanized humanity, lest all things
Be held unalterable!

5
DECENTRALIZATION CLARIFIED

WHAT IS DECENTRALIZATION? People disenchanted with the size, complexity, power, and frequent destructiveness of modern sociotechnical organizations often propose the idea as a key to social reform: let us take our enormous, overly centralized institutions and break them into smaller, human-sized parts more accessible to our control. As we noted in the previous chapter, the call for decentralization has been a crucial theme among activists seeking an appropriate technology. Similar notions are prominently featured in recent attempts to advance the cause of participatory democracy. Spokesmen of both the far Left and the far Right include plans for decentralization in their lists of desirable political goals, although they have much different institutional targets in mind. Even social-trend spotters with no investment in radical or conservative agendas delight in announcing that "centralized structures are crumbling" and that "the decentralization of America has transformed politics, business, our very culture."[1]

But the enthusiasm of such suggestions is seldom matched by any coherent definition or concrete proposal. "Decentralization" is one of the foggiest, most often abused concepts in political language. For those who think it a cure for the ills of modern society, a crucial first step is to clarify what the idea means.

Center of What?

UNFORTUNATELY, THE WORD "decentralization" is something of a linguistic train wreck. Its prefix "de-" means an undoing or reversal, while its suffix "-ization" indicates a becoming, a process under way. Pointing in two directions simultaneously, the word seems to pull itself apart. In the middle of this contradictory motion is the adjective "central," referring to an unspecified center. Taken in a literal sense, therefore, the term means undoing the process of becoming central. The very unwieldiness of the concept reflects the disorientation of those who seek to use it as a formula for change.

Which centers are the important ones? The question is too seldom asked. Geometrical centers as such, the centers of circles or parcels of land, are not at issue. The meaning of "center" in question is usually that of a geographical or institutional place in which a particular kind of activity or influence is concentrated. Thus, we say that New York City is an important center of cultural activity and that the center of corporate planning for the Ford Motor Company is its board of directors. In all discussions of centralization and decentralization, one needs to begin by specifying which kind of activity or influence is problematic. Otherwise one begins to drift into vague expressions about things being taken away or brought closer to us with no solid sense of what those things are.

Once we know which center or centers we are talking about—centers of coal production in North America, the central budgeting agency in a given city, etc.—there are a number of questions that are helpful to ask about centralization and decentralization. How many centers are there? Where are they located? How much power do they possess? How much cultural diversity and vitality do they exhibit?

Looking at the *number* of centers of social activity, the exact count matters less than judgments about whether or not there are relatively too many, too few, or roughly the right number. In some cases a single center—complete centralization—is the ideal condition and anything more would be a nuisance; almost everyone agrees that there ought to be only one central rule-making body for major league baseball. In many areas of life, however, the more genuine centers, the better; to discover that the number of good symphony orchestras in society is increasing would always be good news. Deciding how many centers are the right number prompts us to think about the nature of the activity or influence at issue. In politics, for example, the spectrum of governments range from dictatorship at one extreme to individualistic anarchism at the other. Most political theories specify the need for governmental centers somewhere between one-person rule and the rule of everyone for himself/herself. Justifying exactly how many centers are best is no easy task; it involves some of the most ancient and difficult controversies in political thought.

A second important question has to do with the *location* of centers. Sometimes this matters with regard to a particular geographical place where an activity is centered, for example, the Vatican, Washington, D.C., or Hollywood. Thus, early in their history several states on the East Coast took care to establish their capitals at inland locations, largely to protect farming interests against commercial, coastal traders. At issue was the relationship between geographical places and political influence. More often, however, the crucial matter is the location of centers within an institutional setting. Robert A. Caro explains of Robert Moses' reign over public works in New York, "He centralized in his person and in his projects all those forces in the city that in theory have little to do with the decision-making process in the city's government but in reality have everything to do with it."[2] As Moses moved, so moved an important center of control. But whether a center exists in geographical or social space, what counts most is one's position relative to it. Is it near or far away? Is it accessible or inaccessible? The physical center of something may be literally right next door to us and yet completely beyond reach.

Another significant issue concerning centers has to do with their *power*, especially their relative power as compared to other centers of a similar kind. For our purposes here, power can be understood as the ability of persons or social groups to accomplish their goals. Disputes about centralization and decentralization frequently hinge on the issue of who has how much social, economic, or political power and whether or not the exercise of such power is legitimate. Advocates of centralization often point to advantages of efficiency and superior control that may come from placing power in relatively few hands. In contrast, decentralists argue that from a practical and moral point of view, power is best used when it is widely dispersed.

A fourth question about centers has to do with their *diversity* and *vitality*. Those who stress decentralization as a positive good are often concerned about ensuring the liveliness of modern culture, a condition threatened by the dreary, uniform products that so often emanate from centralized organizations. One reason to prefer a large number and variety of centers for a given activity is that they may be more imaginative and creative than one or just a few centers. This is much different from saying that a large number would have more power. For example, in policy discussions about public funding for the arts there is often a choice of supporting musicians, dance companies, museums, and artists in a few large cities or of using the same money to support the arts and crafts in local communities across the country. It may be, in fact, that geographical centralization of the arts in a few large cities may actually improve the quality of the very best compositions, paintings, films, and other works a society produces. But the decentralist position argues that even greater creativity can be fostered by enabling numerous, small, diverse centers of culture to flourish.

I offer these questions about the number, location, power, diversity, and vitality of centers as clarifying themes in place of a definition. Here is an instance, I believe, in which the attempt to provide a fixed definition of a concept will not help very much. "Decentralization" is simply not a term whose meaning is well established in everyday speech. Rather than arbitrarily impose a definition, it seems best for the time being to notice the meanings people wish to express when they employ this word. Once we have done that, we can begin to notice who cares about the idea and why.

When Centers Matter

THE TERMS "centralized," "decentralized," "centralization," and "decentralization" first appeared in the English language in the early and middle nineteenth century, referring at that time to the relative tightness or looseness of political authority.[3] Over many decades the word "decentralized" has become a fairly common descriptive term in political speech. Paul Goodman's formulation written in the early 1960s captures one basic idea: "Decentralizing is increasing the numbers of centers of decision-making and the number of initiators of policy; increasing the awareness of the whole function in which they are involved; and establishing as much face-to-face association with decision-makers as possible."[4]

Goodman's formulation echoes the beliefs of generations of direct democrats, libertarians, and anarchists that ordinary people are perfectly capable of making decisions and acting for themselves. As a species of modern political ideology, therefore, decentralism stresses the need for a greater number of centers of genuine social and political policy making. The New England town meeting, Spanish anarchist communes, and political practices of the sans-culottes in the French Revolution are instances of decentralism of this sort; they express a desire to involve citizens in public deliberations through direct democratic roles. In contrast, where suspicion of widespread citizen participation prevails, other political structures are invented. Eighteenth-century American federalists succeeded in establishing a structured balance (some might even call it decentralized) between the state and national governments, but they were horrified at the thought that the common people might initiate policy through face-to-face relationships. The U.S. Constitution reflects this fear: its governmental institutions invest far greater authority in a few remote, central law-making bodies than the Articles of Confederation that preceded it.

But centers of decision making are not the only ones that matter in politics, since the making of decisions is not the same as carrying them out. Means of administration, management, regulation, law enforcement, and so forth, constitute another area in which questions about centralization and decentralization arise. What one feels about the where's and how's of decision making may have little to do with one's views on administration. It is possible to insist that a policy issue be settled in a broadly based, grass-roots political manner and at the same time demand that the policy chosen be enforced by a single, powerful national agency.

Historically speaking, those who have chosen to use the term "decentralization" to represent a positive social goal, rather than as a neutral descriptive term, have been partisans of certain factions within the political Left. Asserting the right of people to exercise decision making and administrative authority directly, their concern has been how to accomplish thoroughgoing reform, even revolution, while avoiding the pitfalls of concentrating power in the state. At the turn of the century the writings of Peter Kropotkin, especially *Mutual Aid, a Factor in Evolution* and *Fields, Factories and Workshops*, spelled out a vision of an anarchist social order in which the state has been abolished and all political power is held by small local communes organized in a loosely linked federation.[5] In our own time activist and theorist Murray Bookchin has reformulated classic anarchist notions in a lively series of books, essays, and speeches.[6] Although decentralist ideas have never won widespread following, thinkers in this tradition have helped leaven political thought by imagining a society organized on principles radically different from the ones modern nation states have followed.

A particularly clear, systematic model for a decentralized society was proposed in the 1920s by British social theorist G. D. H. Cole and his colleagues in the Guild Socialist movement. "The essence of the Guild Socialist attitude," Cole explained, "lies in the belief that Society ought to be so organized as to afford the greatest possible opportunity for individual and collective self-expression to all its members, and that this involves and implies the extension of positive self-government through all its parts."[7] To realize this ideal, the base of authority and political representation was to be located in functionally organized local "guilds" within various industries, for example, coal mining, railroads, and the building trades. Each guild would elect representatives to deliberate on broader national issues connected to the guild's specific economic function. A wide range of local, democratically elected civic councils would manage public services and cultural affairs. The state was to remain as a nominal presence, with its power drastically fragmented and curtailed.

Decentralist writers and social movements of the nineteenth century have vehemently opposed the centralization of state power in capitalist societies. But they have reserved special wrath for any similar form that threatened to take root within socialism. From this point of view, centralism is the indelible stain that soils the hopes of socialist humanism. Thus, Cole noticed the Russian model of state socialism and warned against it. "If the Guilds are to revive craftsmanship and pleasure in work well done; if they are to produce quality as well as quantity, and to be ever keen to devise new methods and utilize every fresh discovery of science without loss of tradition; if they are to breed free men capable of being good citizens both in industry and in every aspect of communal life; if they are to keep alive the motive of free service—they must at all costs shun centralization." Cole was confident that a gradual, peaceful adoption of the path he suggested would avoid this evil. "Men freed from the double centralized autocracy of capitalist trust and capitalist State," he argued, "are not likely to make for themselves a new industrial Leviathan."[8]

Many of the issues about political institutions that have fascinated decentralist thinkers appear in a somewhat different light when we consider the material and social structures of modern technology. The ownership, initial source, and conditions of production, distribution, consumption, or use of goods and services can all be described as more-or-less centralized or decentralized. Thus, we can take particular items of value—iron, wood, bananas, refrigerators—and trace their life histories through the various centers that affect their movement from raw material to finished product to eventual waste. Are there many sources for the things we employ or just a few? Are there many producers and distributors or a limited number? How close are we to the sources we rely upon? Who exercises control over them and in what manner? We can also look at the sociotechnical systems that provide services of various kinds—communications, transportation, information handling, waste disposal, medical care, and the like—and notice how the existing structures of such systems embody answers to questions about the number, location, power, and vitality of centers.

In material culture, as in politics, the relative importance of centers changes greatly with the specific activity in question and its setting. One way of testing whether or not the concept "decentralization" is meaningful is to employ it in a particular situation and see if it makes sense. If we observe, for example, that the Exxon Corporation controls roughly ten percent of the petroleum market in the United States, does that justify the conclusion that the oil industry is decentralized? Of course not. The fact that the petroleum business contains a number of different producers, each with a fraction of total sales, does not alter the significant fact that most of these producers are huge, extremely powerful, and centrally controlled within themselves. To say that Exxon, always near the top of the *Fortune* 500 companies, represents a decentralized way of doing things is patently absurd. If, in contrast, I learned that no Chinese restaurant in Portland, Oregon, had more than ten percent of Chinese food sales, I would probably be prepared to say that it was a decentralized business unless I discovered that all such establishments in that city were owned and managed by the same person. What seems a manifestly bad measure of the relative strength of centers in one context turns out to be a pretty good measure in another.

The difficulty in specifying exactly when and how centers matter to us raises some knotty problems for those who suggest decentralizing technology as a variety of social reform. Suppose, for example, that a great number of people began to take seriously the prospect already mentioned several times in this book: that decentralized energy offers the promise of a better future. What could that possibly mean? My arguments and observations so far should indicate why the answer is not a simple one. It might mean that the original sources of energy—coal, wood, solar energy, and so forth—would somehow become more numerous and accessible. It could mean that the kinds of apparatus that carry and transform the energy to make it usable would become more readily available. It could mean that such equipment would be produced, distributed, and marketed by small, local concerns. It might also mean that we would decentralize social policy decision making and administration with respect to energy. Or it might mean some combination of these measures.

If most households in the United States began to employ solar energy for space and water heating, in what sense would that be a decentralized system? One can envision a situation in which solar devices were manufactured by General Electric, distributed by Sears, purchased by consumers, and installed by local plumbers; this is, in fact, one of the more probable of the soft energy paths. While it would enable households to produce their own energy on site without being perpetually hooked up to centrally organized sources of supply, it would also replicate other long-standing centralist patterns in society.

Whether or not we care about the relative number, location, vitality, and power of centers of activity in technology clearly changes with time. To understand recent interest in such matters as decentralized energy or the alleged decentralizing qualities of microcomputers, one has to examine yesterday's lack of interest in such things. Over the last century or so in this country, the production and distribution of many different kinds of commodities and services have been placed under the central dominion of very large private, public, and semipublic organizations. Justified as the most efficient and productive way of doing things, the tendency has been to allow corporations, public utilities, public authorities, and government agencies to manage communications, energy, manufacturing, transportation, and other areas of material culture. Very often, of course, the centers of control that we see today arose by crushing or absorbing their competitors. The large chain stores that first appeared in the last half of the nineteenth century succeeded over bitter opposition of smaller retailers. What began as a relatively decentralized automobile industry in America (in the sense that there were many independent producers) was eventually brought under the control of three large firms. The victors became brand names familiar to this day, while the vanquished are remembered as obscure historical footnotes.

While each specific area of production and distribution has its own history and distinctive mode of organization, the overall pattern is clear. The social history of modern technology shows a tendency—perhaps better termed a strategy—to reduce the number of centers at which action is initiated and control is exercised.

Significantly, the same was not true of the everyday consumption and use of things. In that domain individuals remained the centers of activity. Noting this fact, a number of observers have argued that the automobile is an extremely decentralized form of transportation. That is true as regards the ownership and use of automobiles, although the term "atomistic" is a more accurate way to describe what we observe than "decentralized." However, if we look at the production, marketing, and servicing of the vehicles as well as the building and maintenance of the roads they run on, it is a much different story.

For most of the twentieth century the prospect of unhindered personal consumption and use of goods seemed to make most people happy; in fact, that is what "freedom" began to mean in the eyes of many. It did not matter that General Motors had become so large and powerful, since that meant that the company produced an inexpensive, reliable car for our weekend spin in the country. It did not matter how electricity was made or carried over a vast electric grid as long as you could flick a switch and have the lights go on. It did not matter exactly what Kellogg was putting in the cornflakes as long as the cereal tasted good and seemed to contribute to strong bones and healthy bodies. A primary source of legitimacy for many of the systems that form the heart of the technological society was that consumption was still centered in the individual. Any notion that ordinary people might want to have control over production or have a say in decisions beyond those of the immediate enjoyment of goods and services seemed out of the question.

Alienation from Centers

THE TENDENCY to remove the control over production from centers of everyday life—the home, neighborhood, community, and workplace—has become firmly fixed in both technical hardware and organizational practices of modern society. The designs of all kinds of technologies—electrical power plants, water systems, highways, and vehicles that run on them, the machines and chemicals of agribusiness farming, supermarkets and the products they stock, television, radio, computers, and many others—embody centralization in material form. Artifacts in common use presuppose relatively few centers of production and distribution. In addition, many of our daily activities rely upon systems that we do not make, control, or know how to repair when they break down. It is possible to imagine alternatives to this state of affairs, different ways of structuring the physical and social dimensions of modern material culture. But given the direction that technological "progress" has followed, people find themselves dependent upon a great many large, complex systems whose centers are, for all practical purposes, beyond their power to influence.

The renewed interest in issues of centralization and decentralization during the 1960s and 1970s mirrored a growing alienation with the organizational forms of both politics and technology. Groups of the New Left emphasized distinctly political concerns: participatory democracy and worker self-management as crucial steps in political reform. After a time, however, the topic began to attract a broader public, citizens worried about vexing problems in the workings of major technology-based institutions. Environmental pollution, electrical power blackouts, rising petroleum costs, gasoline shortages, the hazards of commonly used drugs and chemicals, and the possibility of nuclear power plant accidents combined to create suspicion about things that had not previously seemed troubling. Reassurances from the public relations campaigns of large, high-tech corporations—AT&T's slogan that "The System is the Solution" and Monsanto's vacuous reminder that "Without chemicals, life itself would be impossible"—began to seem less salutary than chilling. After the near meltdown at the Three Mile Island nuclear power plant, officials of the nuclear power industry offered the same bizarre consolation that a few years earlier had excused the turmoil of the Watergate scandal, "The system works." Yes, it certainly does. But many began to ask: Have we placed too much power at the disposal of organizations that are insensitive and unreliable?

Alienation from the organized structures of politics and technology reflect some long-standing social habits. A widely held notion in the twentieth century is that political life is little more than getting other people to do things for you. One delegates power and authority to representatives, to bureaucrats, to the President, or other such distant persons to get the business of government out of your way. These are not matters for which ordinary citizens want to be responsible. People watch public events on television, sense that these matters are beyond their reach, and then complain that the "government" is getting too large and powerful. Americans have long flattered themselves by calling this way of doing business "democracy" and "self-government" (of all things). Altogether similar attitudes characterize most of our dealings with organized centers of material production and distribution. Here also we are inclined to say, "Let someone else take care of it." The prevailing sentiments are those of apathy and dissociation. It is only when things go wrong that we begin to question the underlying organizational structures and then only briefly.

A recurring fantasy in industrial society expects relief from this thoroughgoing estrangement in the coming of a new technological system. In 1924, for example, Joseph K. Hart, a professor of education, extolled the liberation electricity would bring. "Centralization," he wrote, "has claimed everything for a century: the results are apparent on every hand. But the reign of steam approaches its end: a new stage in the industrial revolution comes on. Electric power, breaking away from its servitude to steam, is becoming independent. Electricity is a decentralizing form of power: it runs out over distributing lines and subdivides to all the minutiae of life and need. Working with it, men may feel the thrill of control and freedom once again."[9] What Hart's enthusiastic forecast failed to notice, of course, was that this new energy was centrally generated and centrally controlled by utilities, firms destined to have enormous social power. Indeed, the centralism of the age of steam would seem modest compared to corresponding patterns developed in the age of electricity. Dreams of instant liberation from centralized social control have accompanied virtually every important new technological system introduced during the past century and a half. The emancipation proposed by decentralist philosophers as a deliberate goal requiring long, arduous social struggle has been upheld by technological optimists as a condition to be realized simply by adopting a new gadget. This strange mania, as we will see in the next chapter, is alive and well among those who celebrate the advent of the computer revolution.

Of course, it is never that easy. Given the deeply entrenched patterns of our society, any significant attempt to decentralize major political and technological institutions would be a drastic undertaking. To decentralize in politics would require that we change many of the rules, public roles, and institutional relationships of government. It would mean that society move to increase the number, accessibility, relative power, vitality, and diversity of local centers of decision making and public administration. That could only happen by overcoming what would surely be powerful resistance to any such policy. It would require something of a revolution. Similarly, to decentralize technology would mean redesigning and replacing much of our existing hardware and reforming the ways our technologies are managed. One can imagine many different forms these changes might take. But in either the technical or political sphere (or both) any significant move to decentralize would amount to retro-fitting our whole society, since centralized institutions have become the norm.

In Kropotkin's or G. D. H. Cole's time it was still possible to imagine an entire modern social order based upon small-scale, directly democratic, widely dispersed centers of authority. Industrial society had not yet achieved its mature form; it was thinkable that decentralist alternatives might be feasible alternatives on a broad scale. Today, however, ideas of decentralization usually play a much different role, an expression of the faint hope one may still create institutions here and there that allow ordinary folks some small measure of autonomy. No longer primarily a demand for radical social reconstruction, the plea to decentralize often means: If there is any choice in the matter, let us place greater faith in people's ability to make plans, shape policies, and manage their own public affairs. Rather than force all social transactions into the iron vise of bureaucratic and corporatistic megastructures, let's create a few organizational forms that are more flexible, more forgiving. The idea of decentralization contains the hope that people may yet overcome paltry, dis-associated political roles—taxpayer, consumer, representative sample in the Gallup poll—to constitute themselves as new centers where ideas are hatched and action taken.

6
MYTHINFORMATION

Computer power to the people is essential to the realization of a
future in which most citizens are informed about, and interested and involved in, the processes of government.

J. C. R. Licklider

IN NINETEENTH-CENTURY Europe a recurring ceremonial
gesture signaled the progress of popular uprisings. At the point
at which it seemed that forces of disruption in the streets were
sufficiently powerful to overthrow monarchical authority, a
prominent rebel leader would go to the parliament or city hall
to "proclaim the republic." This was an indication to friend
and foe alike that a revolution was prepared to take its work
seriously, to seize power and begin governing in a way that guaranteed political representation to all the people. Subsequent
events, of course, did not always match these grand hopes; on
occasion the revolutionaries were thwarted in their ambitions
and reactionary governments regained control. Nevertheless,
what a glorious moment when the republic was declared! Here,
if only briefly, was the promise of a new order—an age of equality, justice, and emancipation of humankind.

A somewhat similar gesture has become a standard feature in
contemporary writings on computers and society. In countless
books, magazine articles, and media specials some intrepid soul
steps forth to proclaim "the revolution." Often it is called simply "the computer revolution"; my brief inspection of a library
catalogue revealed three books with exactly that title published
since 1962.[1] Other popular variants include the "information
revolution," "microelectronics revolution," and "network revo-

lution." But whatever its label, the message is usually the same. The use of computers and advanced communications technologies is producing a sweeping set of transformations in every corner of social life. An informal consensus among computer scientists, social scientists, and journalists affirms the term "revolution" as the concept best suited to describe these events. "We are all very privileged," a noted computer scientist declares, "to be in this great Information Revolution in which the computer is going to affect us very profoundly, probably more so than the Industrial Revolution."[2] A well-known sociologist writes, "This revolution in the organization and processing of information and knowledge, in which the computer plays a central role, has as its context the development of what I have called the post-industrial society."[3] At frequent intervals during the past dozen years, garish cover stories in *Time* and *Newsweek* have repeated this story, climaxed by *Time*'s selection of the computer as its "Man of the Year" for 1982.

Of course, the same society now said to be undergoing a computer revolution has long since gotten used to "revolutions" in laundry detergents, underarm deodorants, floor waxes, and other consumer products. Exhausted in Madison Avenue advertising slogans, the image has lost much of its punch. Those who employ it to talk about computers and society, however, appear to be making much more serious claims. They offer a powerful metaphor, one that invites us to compare the kind of disruptions seen in political revolutions to the changes we see happening around computer information systems. Let us take that invitation seriously and see where it leads.

A Metaphor Explored

SUPPOSE THAT we were looking at a revolution in a Third World country, the revolution of the Sandinistas in Nicaragua, for example. We would want to begin by studying the fundamental goals of the revolution. Is this a movement truly committed to social justice? Does it seek to uphold a valid ideal of human freedom? Does it aspire to a system of democratic rule? Answers to those questions would help us decide whether or not this is a revolution worthy of our endorsement. By the same token, we would want to ask about the means the revolutionaries had chosen to pursue their goals. Having succeeded in armed struggle, how will they manage violence and military force once they gain control? A reasonable person would also want to learn something of the structure of institutional authority that the revolution will try to create. Will there be frequent, open elections? What systems of decision making, administration, and law enforcement will be put to work? Coming to terms with its proposed ends and means, a sympathetic observer could then watch the revolution unfold, noticing whether or not it remained true to its professed purposes and how well it succeeded in its reforms.

Most dedicated revolutionaries of the modern age have been willing to supply coherent public answers to questions of this sort. It is not unreasonable to expect, therefore, that something like these issues must have engaged those who so eagerly use the metaphor "revolution" to describe and celebrate the advent of computerization. Unfortunately, this is not the case. Books, articles, and media specials aimed at a popular audience are usually content to depict the dazzling magnitude of technical innovations and social effects. Written as if by some universally accepted format, such accounts describe scores of new computer products and processes, announce the enormous dollar value of the growing computer and communications industry, survey the expanding uses of computers in offices, factories, schools, and homes, and offer good news from research and development laboratories about the great promise of the next generation of computing devices. Along with this one reads of the many "impacts" that computerization is going to have on every sphere of life. Professionals in widely separate fields—doctors, lawyers, corporate managers, and scientists—comment on the changes computers have brought to their work. Home consumers give testimonials explaining how personal computers are helping educate their children, prepare their income tax forms, and file their recipes. On occasion, this generally happy story will include reports on people left unemployed in occupations underminded by automation. Almost always, following this formula, there will be an obligatory sentence or two of criticism of the computer culture solicited from a technically qualified spokesman, an attempt to add balance to an otherwise totally sanguine outlook.

Unfortunately, the prevalence of such superficial, unreflective descriptions and forecasts about computerization cannot be attributed solely to hasty journalism. Some of the most prestigious journals of the scientific community echo the claim that a revolution is in the works.[4] A well-known computer scientist has announced unabashedly that "revolution, transformation and salvation are all to be carried out."[5] It is true that more serious approaches to the study of computers and society can be found in scholarly publications. A number of social scientists, computer scientists, and philosophers have begun to explore important issues about how computerization works and what developments, positive and negative, it is likely to bring to society.[6] But such careful, critical studies are by no means the ones most influential in shaping public attitudes about the world of microelectronics. An editor at a New York publishing house stated the norm, "People want to know what's new with computer technology. They don't want to know what could go wrong."[7]

It seems all but impossible for computer enthusiasts to examine critically the *ends* that might guide the world-shaking developments they anticipate. They employ the metaphor of revolution for one purpose only—to suggest a drastic upheaval, one that people ought to welcome as good news. It never occurs to them to investigate the idea or its meaning any further.

One might suppose, for example, that a revolution of this type would involve a significant shift in the locus of power; after all, that is exactly what one expects in revolutions of a political kind. Is something similar going to happen in this instance?

One might also ask whether or not this revolution will be strongly committed, as revolutions often are, to a particular set of social ideals. If so, what are the ideals that matter? Where can we see them argued?

To mention revolution also brings to mind the relationships of different social classes. Will the computer revolution bring about the victory of one class over another? Will it be the occasion for a realignment of class loyalties?

In the busy world of computer science, computer engineering, and computer marketing such questions seldom come up. Those actively engaged in promoting the transformation— hardware and software engineers, managers of microelectronics firms, computer salesmen, and the like—are busy pursuing their own ends: profits, market share, handsome salaries, the intrinsic joy of invention, the intellectual rewards of programming, and the pleasures of owning and using powerful machines. But the sheer dynamism of technical and economic activity in the computer industry evidently leaves its members little time to ponder the historical significance of their own activity. They must struggle to keep current, to be on the crest of the next wave as it breaks. As one member of Data General's Eagle computer project describes it, the prevailing spirit resembles a game of pinball. "You win one game, you get to play another. You win with this machine, you get to build the next."[8] The process has its own inertia.

Hence, one looks in vain to the movers and shakers in computer fields for the qualities of social and political insight that characterized revolutionaries of the past. Too busy. Cromwell, Jefferson, Robespierre, Lenin, and Mao were able to reflect upon the world historical events in which they played a role. Public pronouncements by the likes of Robert Noyce, Marvin Minsky, Edward Feigenbaum, and Steven Jobs show no similar wisdom about the transformations they so actively help to create. By and large the computer revolution is conspicuously silent about its own ends.

Good Console, Good Network, Good Computer

MY CONCERN for the political meaning of revolution in this setting may seem somewhat misleading, even perverse. A much better point of reference might be the technical "revolutions" and associated social upheavals of the past, the industrial revolution in particular. If the enthusiasts of computerization had readily taken up this comparison, studying earlier historical periods for similarities and differences in patterns of technological innovation, capital formation, employment, social change, and the like, then it would be clear that I had chosen the wrong application of this metaphor. But, in fact, no well-developed comparisons of that kind are to be found in the writings on the computer revolution. A consistently ahistorical viewpoint prevails. What one often finds emphasized, however, is a vision of drastically altered social and political conditions, a future upheld as both desirable and, in all likelihood, inevitable. Politics, in other words, is not a secondary concern for many computer enthusiasts; it is a crucial, albeit thoughtless, part of their message.

We are, according to a fairly standard account, moving into an age characterized by the overwhelming dominance of electronic information systems in all areas of human practice. Industrial society, which depended upon material production for its livelihood, is rapidly being supplanted by a society of information services that will enable people to satisfy their economic and social needs. What water- and steam-powered machines were to the industrial age, the computer will be to the era now dawning. Ever-expanding technical capacities in computation and communications will make possible a universal, instantaneous access to enormous quantities of valuable information. As these technologies become less and less expensive and more and more convenient, all the people of the world, not just the wealthy, will be able to use the wonderful services that information machines make available. Gradually, existing differences between rich and poor, advantaged and disadvantaged, will begin to evaporate. Widespread access to computers will produce a society more democratic, egalitarian, and richly diverse than any previously known. Because "knowledge is power," because electronic information will spread knowledge into every corner of world society, political influence will be much more widely shared. With the personal computer serving as the great equalizer, rule by centralized authority and social class dominance will gradually fade away. The marvelous promise of a "global village" will be fulfilled in a worldwide burst of human creativity.

A sampling from recent writings on the information society illustrates these grand expectations.

> The world is entering a new period. The wealth of nations, which depended upon land, labor, and capital during its agricultural and industrial phases—depended upon natural resources, the accumulation of money, and even upon weaponry— will come in the future to depend upon information, knowledge and intelligence.[9]
>
> * * *
>
> The electronic revolution will not do away with work, but it does hold out some promises: Most boring jobs can be done by machines; lengthy commuting can be avoided; we can have enough leisure to follow interesting pursuits outside our work; environmental destruction can be avoided; the opportunities for personal creativity will be unlimited.[10]

Long lists of specific services spell out the utopian promise of this new age: interactive television, electronic funds transfer, computer-aided instruction, customized news service, electronic magazines, electronic mail, computer teleconferencing, on-line stock market and weather reports, computerized Yellow Pages, shopping via home computer, and so forth. All of it is supposed to add up to a cultural renaissance.

Whatever the limits to growth in other fields, there are no limits near in telecommunications and electronic technology. There are no limits near in the consumption of information, the growth of culture, or the development of the human mind.[11]

* * *

Computer-based communications can be used to make human lives richer and freer, by enabling persons to have access to vast stores of information, other "human resources," and opportunities for work and socializing on a more flexible, cheaper and convenient basis than ever before.[12]

* * *

When such systems become widespread, potentially intense communications networks among geographically dispersed persons will become actualized. We will become Network Nation, exchanging vast amounts of information and social and emotional communications with colleagues, friends and "strangers" who share similar interests, who are spread all over the nation.[13]

* * *

A rich diversity of subcultures will be fostered by computer-based communications systems. Social, political, technical changes will produce conditions likely to lead to the formation of groups with their own distinctive sets of values, activities, language and dress.[14]

According to this view, the computer revolution will, by its sheer momentum, eliminate many of the ills that have vexed political society since the beginning of time. Inequalities of wealth and privilege will gradually fade away. One writer predicts that computer networks will "offer major opportunities to disadvantaged groups to acquire the skills and social ties they need to become full members of society."[15] Another looks forward to "a revolutionary network where each node is equal in power to all others."[16] Information will become the dominant form of wealth. Because it can flow so quickly, so freely through computer networks, it will not, in this interpretation, cause the kinds of stratification associated with traditional forms of property. Obnoxious forms of social organization will also be replaced. "The computer will smash the pyramid," one best-selling book proclaims. "We created the hierarchical, pyramidal, managerial system because we needed it to keep track of people and things people did; with the computer to keep track, we can restructure our institutions horizontally."[17] Thus, the proliferation of electronic information will generate a leveling effect to

104

surpass the dreams of history's great social reformers.

The same viewpoint holds that the prospects for participatory democracy have never been brighter. According to one group of social scientists, "The form of democracy found in the ancient Greek city-state, the Israeli kibbutz, and the New England town meeting, which gave every citizen the opportunity to directly participate in the political process, has become impractical in America's mass society. But this need not be the case. The technological means exist through which millions of people can enter into dialogue with one another and with their representatives, and can form the authentic consensus essential for democracy."[18]

Computer scientist J. C. R. Licklider of the Massachusetts Institute of Technology is one advocate especially hopeful about a revitalization of the democratic process. He looks forward to "an information environment that would give politics greater depth and dimension than it now has." Home computer consoles and television sets would be linked together in a massive network. "The political process would essentially be a giant teleconference, and a campaign would be a months-long series of communications among candidates, propagandists, commentators, political action groups and voters." An arrangement of this kind would, in his view, encourage a more open, comprehensive examination of both issues and candidates. "The information revolution," he exclaims, "is bringing with it a key that may open the door to a new era of involvement and participation. The key is the self-motivating exhilaration that accompanies truly effective interaction with information through a good console through a good network to a good computer."[19] It is, in short, a democracy of machines.

Taken as a whole, beliefs of this kind constitute what I would call mythinformation: the almost religious conviction that a widespread adoption of computers and communications systems along with easy access to electronic information will automatically produce a better world for human living. It is a peculiar form of enthusiasm that characterizes social fashions of the latter decades of the twentieth century. Many people who have grown cynical or discouraged about other aspects of social life are completely enthralled by the supposed redemptive qualities of computers and telecommunications. Writing of the "fifth generation" supercomputers, Japanese author Yoneji Masuda rhapsodically predicts "freedom for each of us to set individual goals of self-realization and then perhaps a worldwide religious renaissance, characterized not by a belief in a supernatural god, but rather by awe and humility in the presence of the collective human spirit and its wisdom, humanity living in a symbolic tranquility with the planet we have found ourselves upon, regulated by a new set of global ethics."[20]

It is not uncommon for the advent of a new technology to provide an occasion for flights of utopian fancy. During the last two centuries the factory system, railroads, telephone, electricity, automobile, airplane, radio, television, and nuclear power have all figured prominently in the belief that a new and glorious age was about to begin. But even within the great tradition of optimistic technophilia, current dreams of a "computer age" stand out as exaggerated and unrealistic. Because they have such a broad appeal, because they overshadow other ways of looking at the matter, these notions deserve closer inspection.

The Great Equalizer

As is generally true of a myth, the story contains elements of truth. What were once industrial societies are being transformed into service economies, a trend that emerges as more material production shifts to developing countries where labor costs are low and business tax breaks lucrative. At the same time that industrialization takes hold in less-developed nations of the world, deindustrialization is gradually altering the economies of North America and Europe. Some of the service industries central to this pattern are ones that depend upon highly sophisticated computer and communications systems. But this does not mean that future employment possibilities will flow largely from the microelectronics industry and information services. A number of studies, including those of the U.S. Bureau of Labor Statistics, suggest that the vast majority of new jobs will come in menial service occupations paying relatively low wages.[21] As robots and computer software absorb an increasing share of factory and office tasks, the "information society" will offer plenty of opportunities for janitors, hospital orderlies, and fast-food waiters.

The computer romantics are also correct in noting that computerization alters relationships of social power and control, although they misrepresent the direction this development is likely to take. Those who stand to benefit most obviously are large transnational business corporations. While their "global reach" does not arise solely from the application of information technologies, such organizations are uniquely situated to exploit the efficiency, productivity, command, and control the new electronics make available. Other notable beneficiaries of the systematic use of vast amounts of digitized information are public bureaucracies, intelligence agencies, and an ever-expanding military, organizations that would operate less effectively at their present scale were it not for the use of computer power. Ordinary people are, of course, strongly affected by the workings of these organizations and by the rapid spread of new electronic systems in banking, insurance, taxation, factory and office work, home entertainment, and the like. They are also counted upon to be eager buyers of hardware, software, and communications services as computer products reach the consumer market.

But where in all of this motion do we see increased democratization? Social equality? The dawn of a cultural renaissance? Current developments in the information age suggest an increase in power by those who already had a great deal of power, an enhanced centralization of control by those already prepared for control, an augmentation of wealth by the already wealthy. Far from demonstrating a revolution in patterns of social and political influence, empirical studies of computers and social change usually show powerful groups adapting computerized methods to retain control.[22] That is not surprising. Those best situated to take advantage of the power of a new technology are often those previously well situated by dint of wealth, social standing, and institutional position. Thus, if there is to be a computer revolution, the best guess is that it will have a distinctly conservative character.

Granted, such prominent trends could be altered. It is possible that a society strongly rooted in computer and telecommunications systems could be one in which participatory democracy, decentralized political control, and social equality are fully realized. Progress of that kind would have to occur as the result of that society's concerted efforts to overcome many difficult obstacles to achieve those ends. Computer enthusiasts, however, seldom propose deliberate action of that kind. Instead, they strongly suggest that the good society will be realized as a side effect, a spin-off from the vast proliferation of computing devices. There is evidently no need to try to shape the institutions of the information age in ways that maximize human freedom while placing limits upon concentrations of power.

For those willing to wait passively while the computer revolution takes its course, technological determinism ceases to be mere theory and becomes an ideal: a desire to embrace conditions brought on by technological change without judging them in advance. There is nothing new in this disposition. Computer romanticism is merely the latest version of the nineteenth- and twentieth-century faith we noted earlier, one that has always expected to generate freedom, democracy, and justice through sheer material abundance. Thus there is no need for serious inquiry into the appropriate design of new institutions or the distribution of rewards and burdens. As long as the economy is growing and the machinery in good working order, the rest will take care of itself. In previous versions of this homespun conviction, the abundant (and therefore democratic) society was manifest by a limitless supply of houses, appliances, and consumer goods.[23] Now "access to information" and "access to computers" have moved to the top of the list.

The political arguments of computer romantics draw upon a number of key assumptions: (1) people are bereft of information; (2) information is knowledge; (3) knowledge is power; and (4) increasing access to information enhances democracy and equalizes social power. Taken as separate assertions and in combination, these beliefs provide a woefully distorted picture of the role of electronic systems in social life.

Is it true that people face serious shortages of information? To read the literature on the computer revolution one would suppose this to be a problem on a par with the energy crisis of the 1970s. The persuasiveness of this notion borrows from our sense that literacy, education, knowledge, well-informed minds, and the widespread availability of tools of inquiry are unquestionable social goods, and that, in contrast, illiteracy, inadequate education, ignorance, and forced restrictions upon knowledge are among history's worst evils. Thus, it appears superficially plausible that a world rewired to connect human beings to vast data banks and communications systems would be a progressive step. Information shortage would be remedied in much the same way that developing a new fuel supply might solve an energy crisis.

Alas, the idea is entirely faulty. It mistakes sheer supply of information with an educated ability to gain knowledge and act effectively based on that knowledge. In many parts of the world that ability is sadly lacking. Even some highly developed societies still contain chronic inequalities in the distribution of good education and basic intellectual skills. The U.S. Army, for instance, must now reject or dismiss a fairly high percentage of the young men and women it recruits because they simply cannot read military manuals. It is no doubt true of these recruits that they have a great deal of information about the world—information from their life experiences, schooling, the mass media, and so forth. What makes them "functionally illiterate" is that they have not learned to translate this information into a mastery of practical skills.

If the solution to problems of illiteracy and poor education were a question of information supply alone, then the best policy might be to increase the number of well-stocked libraries, making sure they were built in places where libraries do not presently exist. Of course, that would do little good in itself unless people are sufficiently well educated to use those libraries to broaden their knowledge and understanding. Computer enthusiasts however, are not noted for their calls to increase support of public libraries and schools. It is *electronic information* carried by *networks* they uphold as crucial. Here is a case in which an obsession with a particular kind of technology causes one to disregard what are obvious problems and clear remedies. While it is true that systems of computation and communications, intelligently structured and wisely applied, might help a society raise its standards of literacy, education, and general knowledgeability, to look to those instruments first while ignoring how to enlighten and invigorate a human mind is pure foolishness.

"As everybody knows, knowledge is power."[24] this is an attractive idea, but highly misleading. Of course, knowledge employed in particular circumstances can help one act effectively and in that sense enhance one's power. A citrus farmer's knowledge of frost conditions enables him/her to take steps to prevent damage to the crop. A candidate's knowledge of public opinion can be a powerful aid in an election campaign. But surely there is no automatic, positive link between knowledge and power, especially if that means power in a social or political sense. At times knowledge brings merely an enlightened impotence or paralysis. One may know exactly what to do but lack the wherewithal to act. Of the many conditions that affect the phenomenon of power, knowledge is but one and by no means the

109

most important. Thus, in the history of ideas, arguments that expert knowledge ought to play a special role in politics—the philosopher-kings for Plato, the engineers for Veblen—have always been offered as something contrary to prevailing wisdom. To Plato and Veblen it was obvious that knowledge was *not* power, a situation they hoped to remedy.

An equally serious misconception among computer enthusiasts is the belief that democracy is first and foremost a matter of distributing information. As one particularly flamboyant manifesto exclaims: "There is an explosion of information dispersal in the technology and we think this information has to be shared. All great thinkers about democracy said that the key to democracy is access to information. And now we have a chance to get information into people's hands like never before."[25] Once again such assertions play on our belief that a democratic public ought to be open-minded and well informed. One of the great evils of totalitarian societies is that they dictate what people can know and impose secrecy to restrict freedom. But democracy is not founded solely (or even primarily) upon conditions that affect the availability of information. What distinguishes it from other political forms is a recognition that the people as a whole are capable of self-government and that they have a rightful claim to rule. As a consequence, political society ought to build institutions that allow or even encourage a great latitude of democratic participation. How far a society must go in making political authority and public roles available to ordinary people is a matter of dispute among political theorists. But no serious student of the question would give much credence to the idea that creating a universal gridwork to spread electronic information is, by itself, a democratizing step.

What, then, of the idea that "interaction with information through a good console, through a good network to a good computer" will promote a renewed sense of political involvement and participation? Readers who believe that assertion should contact me about some parcels of land my uncle has for sale in Florida. Relatively low levels of citizen participation prevail in some modern democracies, the United States, for example. There are many reasons for this, many ways a society might try to improve things. Perhaps opportunities to serve in public office or influence public policy are too limited; in that case, broaden the opportunities. Or perhaps choices placed before citizens are so pallid that boredom is a valid response; in that instance, improve the quality of those choices. But it is simply not reasonable to assume that enthusiasm for political activity will be stimulated solely by the introduction of sophisticated information machines.

The role that television plays in modern politics should suggest why this is so. Public participation in voting has steadily declined as television replaced the face-to-face politics of precincts and neighborhoods. Passive monitoring of electronic news and information allows citizens to feel involved while dampening the desire to take an active part. If people begin to rely upon computerized data bases and telecommunications as a primary means of exercising power, it is conceivable that genuine political knowledge based in first-hand experience would vanish altogether. The vitality of democratic politics depends upon people's willingness to act together in pursuit of their common ends. It requires that on occasion members of a community appear before each other in person, speak their minds, deliberate on paths of action, and decide what they will do.[26] This is considerably different from the model now upheld as a breakthrough for democracy: logging onto one's computer, receiving the latest information, and sending back an instantaneous digitized response.

A chapter from recent political history illustrates the strength of direct participation in contast to the politics of electronic information. In 1981 and 1982 two groups of activists set about to do what they could to stop the international nuclear arms race. One of the groups, Ground Zero, chose to rely almost solely upon mass communications to convey its message to the public. Its leaders appeared on morning talk shows and evening news programs on all three major television networks. They followed up with a mass mail solicitation using addresses from a computerized data base. At the same time another group, the Nuclear Weapons Freeze Campaign, began by taking its proposal for a bilateral nuclear freeze to New England town meetings, places where active citizen participation is a long-standing tradition. Winning the endorsement of the idea from a great many town meetings, the Nuclear Freeze group expanded its drive by launching a series of state initiatives. Once again the key was a direct approach to people, this time through thousands of meetings, dinners, and parties held in homes across the country.

The effects of the two movements were strikingly different. After its initial publicity, Ground Zero was largely ignored. It had been an ephemeral exercise in media posturing. The Nuclear Freeze campaign, however, continued to gain influence in the form of increasing public support, successful ballot measures, and an ability to apply pressure upon political officials. Eventually, the latter group did begin to use computerized mailings, television appearances, and the like to advance its cause. But it never forgot the original source of its leverage: people working together for shared ends.

Of all the computer enthusiasts' political ideas, there is none more poignant than the faith that the computer is destined to become a potent equalizer in modern society. Support for this belief is found in the fact that small "personal" computers are becoming more and more powerful, less and less expensive, and ever more simple to use. Obnoxious tendencies associated with the enormous, costly, technically inaccessible computers of the recent past are soon to be overcome. As one writer explains, "The great forces of centralization that characterized mainframe and minicomputer design of that period have now been reversed." This means that "the puny device that sits innocuously on the desktop will, in fact, within a few years, contain enough computing power to become an effective equalizer."[27] Presumably, ordinary citizens equipped with microcomputers will be able to counter the influence of large, computer-based organizations.

Notions of this kind echo beliefs of eighteenth- and nineteenth-century revolutionaries that placing fire arms in the hands of the people was crucial to overthrowing entrenched authority. In the American Revolution, French Revolution, Paris Commune, and Russian Revolution the role of "the people armed" was central to the revolutionary program. As the military defeat of the Paris Commune made clear, however, the fact that the popular forces have guns may not be decisive. In a contest of force against force, the larger, more sophisticated, more ruthless, better equipped competitor often has the upper hand. Hence, the availability of low-cost computing power may move the baseline that defines electronic dimensions of social influence, but it does not necessarily alter the relative balance of power. Using a personal computer makes one no more powerful vis-à-vis, say, the National Security Agency than flying a hang glider establishes a person as a match for the U.S. Air Force.

In sum, the political expectations of computer enthusiasts are seldom more than idle fantasy. Beliefs that widespread use of computers will cause hierarchies to crumble, inequality to tumble, participation to flourish, and centralized power to dissolve simply do not withstand close scrutiny. The formula information = knowledge = power = democracy lacks any real substance. At each point the mistake comes in the conviction that computerization will inevitably move society toward the good life. And no one will have to raise a finger.

Information and Ideology

DESPITE ITS SHORTCOMINGS as political theory, mythinformation is noteworthy as an expressive contemporary ideology. I use the term "ideology" here in a sense common in social science: a set of beliefs that expresses the needs and aspirations of a group, class, culture, or subculture. In this instance the needs and aspirations that matter most are those that stem from operational requirements of highly complex systems in an advanced technological society; the groups most directly involved are those who build, maintain, operate, improve, and market these systems. At a time in which almost all major components of our technological society have come to depend upon the application of large and small computers, it is not surprising that computerization has risen to ideological prominence, an expression of grand hopes and ideals.

What is the "information" so crucial in this odd belief system, the icon now so greatly cherished? We have seen enough to appreciate that the kind of information upheld is not knowledge in the ordinary sense of the term; nor is it understanding, enlightenment, critical thought, timeless wisdom, or the content of a well-educated mind. If one looks carefully at the writings of computer enthusiasts, one finds that information in a particular form and context is offered as a paradigm to inspire emulation. Enormous quantities of data, manipulated within various kinds of electronic media and used to facilitate the transactions of today's large, complex organizations is the model we are urged to embrace. In this context the sheer quantity of information presents a formidable challenge. Modern organizations are continually faced with overload, a flood of data that threatens to become unintelligible to them. Computers provide one way to confront that problem; speed conquers quantity. An equally serious challenge is created by the fact that the varieties of information most crucial to modern organizations are highly time specific. Data on stock market prices, airline traffic, weather conditions, international economic indicators, military intelligence, public opinion poll results, and the like are useful for very short periods of time. Systems that gather, organize, analyze, and utilize electronic data in these areas must be closely tuned to the very latest developments. If one is trading on fast-paced international markets, information about prices an hour old or even a few seconds old may have no value. Information is itself a perishable commodity.

Thus, what looked so puzzling in another context—the urgent "need" for information in a social world filled with many pressing human needs—now becomes transparent. It is, in the first instance, the need of complex human/machine systems threatened with debilitating uncertainties or even breakdown unless continually replenished with up-to-the-minute electronic information about their internal states and operating environments. Rapid information-processing capabilities of modern computers and communications devices are a perfect match for such needs, a marriage made in technological heaven.

But is it sensible to transfer this model, as many evidently wish, to all parts of human life? Must activities, experiences, ideas, and ways of knowing that take a longer time to bear fruit adapt to the speedy processes of digitized information processing? Must education, the arts, politics, sports, home life, and all other forms of social practice be transformed to accommodate it? As one article on the coming of the home computer concludes, "running a household is actually like running a small business. You have to worry about inventory control—of household supplies—and budgeting for school tuition, housekeepers' salaries, and all the rest."[28] The writer argues that these complex, rapidly changing operations require a powerful information-processing capacity to keep them functioning smoothly. One begins to wonder how everyday activities such as running a household were even possible before the advent of microelectronics. This is a case in which the computer is a solution frantically in search of a problem.

In the last analysis, the almost total silence about the ends of the "computer revolution" is filled by a conviction that information processing is something valuable in its own right. Faced with an information explosion that strains the capacities of traditional institutions, society will renovate its structure to accommodate computerized, automated systems in every area of concern. The efficient management of information is revealed as the *telos* of modern society, its greatest mission. It is that fact to which mythinformation adds glory and glitter. People must be convinced that the human burdens of an information age—unemployment, de-skilling, the disruption of many social patterns—are worth bearing. Once again, those who push the plow are told they ride a golden chariot.

Everywhere and Nowhere

HAVING CRITICIZED a point of view, it remains for me to suggest what topics a serious study of computers and politics should pursue. The question is, of course, a very large one. If the long-term consequences of computerization are anything like the ones commonly predicted, they will require a rethinking of many fundamental conditions in social and political life. I will mention three areas of concern.

As people handle an increasing range of their daily activities through electronic instruments—mail, banking, shopping, entertainment, travel plans, and so forth—it becomes technically feasible to monitor these activities to a degree heretofore inconceivable. The availability of digitized footprints of social transactions affords opportunities that contain a menacing aspect. While there has been a great deal written about this problem, most of it deals with the "threat to privacy," the possibility that someone might gain access to information that violates the sanctity of one's personal life. As important as that issue certainly is, it by no means exhausts the potential evils created by electronic data banks and computer matching. The danger extends beyond the private sphere to affect the most basic of public freedoms. Unless steps are taken to prevent it, we may develop systems capable of a perpetual, pervasive, apparently benign surveillance. Confronted with omnipresent, all-seeing data banks, the populace may find passivity and compliance the safest route, avoiding activities that once represented political liberty. As a badge of civic pride a citizen may announce, "I'm not involved in anything a computer would find the least bit interesting."

The evolution of this unhappy state of affairs does not necessarily depend upon the "misuse" of computer systems. The prospect we face is really much more insidious. An age rich in electronic information may achieve wonderful social conveniences at a cost of placing freedom, perhaps inadvertently, in a deep chill.

A thoroughly computerized world is also one bound to alter conditions of human sociability. The point of many applications of microelectronics, after all, is to eliminate social layers that were previously needed to get things done. Computerized bank tellers, for example, have largely done away with small, local branch banks, which were not only ways of doing business, but places where people met, talked, and socialized. The so-called electronic cottage industry, similarly, operates very well without the kinds of human interactions that once characterized office work. Despite greater efficiency, productivity, and convenience, innovations of this kind do away with the reasons people formerly had for being together, working together, acting together. Many practical activities once crucial to even a minimal sense of community life are rendered obsolete. One consequence of these developments is to pare away the kinds of face-to-face contact that once provided important buffers between individuals and organized power. To an increasing extent, people will become even more susceptible to the influence of employers, news media, advertisers, and national political leaders. Where will we find new institutions to balance and mediate such power?

Perhaps the most significant challenge posed by the linking of computers and telecommunications is the prospect that the basic structures of political order will be recast. Worldwide computer, satellite, and communication networks fulfill, in large part, the modern dream of conquering space and time. These systems make possible instantaneous action at any point on the globe without limits imposed by the specific location of the initiating actor. Human beings and human societies, however, have traditionally found their identities within spatial and temporal limits. They have lived, acted, and found meaning in a particular place at a particular time. Developments in microelectronics tend to dissolve these limits, thereby threatening the integrity of social and political forms that depend on them. Aristotle's observation that "man is a political animal" meant in its most literal sense that man is a *polis* animal, a creature naturally suited to live in a particular kind of community within a specific geographical setting, the city-state. Historical experience shows that it is possible for human beings to flourish in political units—kingdoms, empires, nation-states—larger than those the Greeks thought natural. But until recently the crucial conditions created by spatial boundaries of political societies were never in question.

That has changed. Methods pioneered by transnational corporations now make it possible for organizations of enormous size to manage their activities effectively across the surface of the planet. Business units that used to depend upon spatial proximity can now be integrated through complex electronic signals. If it seems convenient to shift operations from one area of the world to another far distant, it can be accomplished with a flick of a switch. Close an office in Sunnyvale; open an office in Singapore. In the recent past corporations have had to demonstrate at least some semblance of commitment to geographically based communities; their public relations often stressed the fact that they were "good neighbors." But in an age in which organizations are located everywhere and nowhere, this commitment easily evaporates. A transnational corporation can play fast and loose with everyone, including the country that is ostensibly its "home." Towns, cities, regions, and whole nations are forced to swallow their pride and negotiate for favors. In that process, political authority is gradually redefined.

Computerization resembles other vast, but largely unconscious experiments in modern social and technological history, experiments of the kind noted in earlier chapters. Following a step-by-step process of instrumental improvements, societies create new institutions, new patterns of behavior, new sensibilities, new contexts for the exercise of power. Calling such changes "revolutionary," we tacitly acknowledge that these are matters that require reflection, possibly even strong public action to ensure that the outcomes are desirable. But the occasions for reflection, debate, and public choice are extremely rare indeed. The important decisions are left in private hands inspired by narrowly focused economic motives. While many recognize that these decisions have profound consequences for our common life, few seem prepared to own up to that fact. Some observers forecast that "the computer revolution" will eventually be guided by new wonders in artificial intelligence. Its present course is influenced by something much more familiar: the absent mind.

III

EXCESS AND LIMIT

7

THE STATE OF
NATURE REVISITED

Seeking a reliable way to judge the works and accomplishments of civilization, each generation finds an obvious point of reference. Nature provides a vivid contrast to human artifice, a source of powerful insights and arguments. Many political theories begin on exactly this note, asserting as an explicit or implicit first premise: this is the natural way; here is the path nature itself sets before us. In the same vein many criticisms of human institutions, including criticisms of technological society, rest on the charge: what we are doing is horribly contrary to nature; we must repair our ways or stand condemned by the most severe of tribunals. Ralph Waldo Emerson's classic essay "Nature" depicts the situation clearly in the context of a man walking into a forest. "The knapsack of custom falls off his back with the first step he takes into these precincts. Here is sanctity which shames our religions, and reality which discredits our heroes. Here we find Nature to be the circumstance which dwarfs every other circumstance, and judges like a god all men that come to her."[1]

Ideas of this kind become more interesting and problematic when expressed as organized doctrine. No artist, no thinker, no political movement, no society has ever rested content with the simple definition of nature as "the totality of all things." From the vast array of natural phenomena, people inevitably select particular features for emphasis, endowing them with great esthetic, moral, and political significance. Anyone who examines the range of meanings that have been attributed to "nature" in Western history must be impressed by their number, diversity, and glaring contradictory implications. For some nature brings conflict, for others harmony; for some it exists as the very essence of reason and order, for others it looms as throbbing irrational passion; for some it is a source of warmth, nourishment, and solace, for others it is a set of awesome, threatening forces. In different historical periods the same symbols of nature carry different, sometimes opposite meanings.[2] Thus, the "wilderness" feared as a haunt of demons by medieval Christians has commonly been regarded as a place of beauty and inspiration in the eyes of industrial society.

To invoke "nature" or "the natural" in discussions about social life is in effect asserting: "This is real. This is trustworthy. I am not making it up." Because natural phenomena existed before human intervention, they have a certain reliability, unlike human artifacts and institutions that are all too often filled with deceit. On those grounds the bastard Edmund in *King Lear* appeals to nature to justify his claims, renouncing those civilized conventions that label him "illegitimate."

> Thou, Nature art my goddess, to thy law
> My services are bound. Wherefore should I
> Stand in the plague of custom, and permit
> The curiosity of nations to deprive me . . .

Shakespeare's play offers the spectacle of political society and its categories dissolving when confronted by uncontrolled natural forces. It was this very solvent that philosophers of the seventeenth and eighteenth centuries employed to discredit existing authority and to prepare the way for the establishment of new regimes. Thus, Thomas Hobbes describes a state of nature so threatening to men, a continuing "war of every one against every one," that any reasonable person would gladly take shelter in a covenant creating the commonwealth.[3]

But what philosophers and social theorists have written about "nature" over the centuries provides adequate grounds for skepticism on the matter. At one point in *The Second Treatise of Government* John Locke addresses the objection some have voiced that "the state of Nature" he describes never existed. He finesses the point by asserting "that all Men are naturally in that State, and remain so, till by their own Consents they make themselves Members of some Politick Society."[4] In other words, a state of nature is logically defined as an event within a set of philosophical constructs. As the writings of Hobbes, Locke, and Rousseau clearly demonstrate, discussions of "natural law," "human nature," and "state of nature" are occasions for the most extravagant theoretical fictions. Nevertheless, because they carry the appearance of reliability, because they offer the promise of universal truth, conceptions of nature have an astonishing power to persuade.

During the last half of the twentieth century the age-old conversation about the meaning of nature has been renewed. Terms that matter now are, for example, "environment," "ecology," and "ecosystem," but the familiar quest to find moral guidance in natural phenomena is similar to that seen in earlier periods. At issue now is the present and future path of modern technological society. Are there any lessons from available knowledge of natural processes that are persuasive enough to cause people to alter the course industrial societies have followed in recent times? Does an awareness of significant features of the "environment" or "ecology" generate imperatives to guide our actions? Do present-day understandings of nature shed new light on the question of limits? Let us examine three contemporary perspectives on environmental policy to see what they discover in nature and what enlightened policies they recommend.

Nature as a Stock of Economic Goods

THE USE of the term "environment" in referring to things in nature is itself testimony to the success of a project that characterizes modern industrial civilization: the control and exploitation of the material world. "Environment" means, literally, "something that surrounds." It suggests that what previous generations had thought of as a target for domination, a frontier "out there" awaiting mastery, is now something that we have moved into and taken over. The "environment" surrounds us in the way a military general is surrounded by countryside he has just conquered. Descartes suggested long ago that we "render ourselves the masters and possessors of nature"[5] and Francis Bacon that we "extend the power and dominion of the human race itself over the universe."[6] This project has succeeded beyond those thinkers' wildest dreams. The extent of the conquest occasionally gives pause to even those who believe that this victory is fully justified. For the value of many things, including market goods, often stems from the fact that they have somehow been left unexploited, unsoiled.

An important segment of today's environmental movement in the United States inherits and extends two strands of public policy that have their origin in Progressive Era politics of the turn of the century: conservation of resources and pollution control. In this tradition certain obvious problems such as deforestation, erosion, and the squandering of minerals, fish, "game," and "scenic" wilderness create challenges for rational husbandry.[7] The unintended, undesirable waste products from industrialization and urbanization present a need for cleanup and preventive measures.[8] But the continuing presence of such problems is no cause to doubt the right of humans to possess and dispose of "natural resources" as they please, no reason to doubt the basic soundness of modern civilization.

There is no felt need, for example, to reexamine older theories that define man's proprietary relationship to natural creation. In Locke's *Second Treatise* we are told of a situation in which men confront parts of nature and transform them into personal possessions. "Whatsoever then he removes out of the State that Nature hath provided, and left it in, he hath mixed his Labour with, and thereby makes it his Property."[9] Significant in Locke's theory is the fact that property is actually created *before* the founding of society. It is only when the state of nature proves inadequate for the protection of life and property that people bind together in a contract establishing civil society and government. We have left the state of nature in order to guarantee safe access to nature as a vast reservoir of useful raw materials. In that move nature is designated, once and for all, as a stock of economic goods.

Unlike the two other viewpoints we will examine, the economic perspective does not depend on drawing lessons from nature or from the human relationship to nature. Intricate artificial models of "the economy" and of "the market" will now do just fine. Still interesting in this context, however, are those "environmental values" that pose a challenge to the penchant to define every situation according to an orthodox format of costs, benefits, supply, demand, and prices. The value of clean air, clean water, dwindling resources, wilderness, and the like is still something to be reckoned with.[10] What one needs is a way of allocating these goods in a rational manner. The orthodox economic means for achieving rational, efficient outcomes in this context is to make sure that society has properly functioning markets that give everything an appropriate price. Hence, the proposals of economists on environmental matters and public policy measures based upon their proposals are often ways of ensuring that environmental values are somehow expressed in hard dollar prices.

To this end, many of the accomplishments of contemporary environmental economics have been attempts to recognize and remedy the shortcomings of the standard neoclassical economic paradigm. Many now appreciate that it will not do to mention "negative externalities" and "market imperfections" and let it go at that. If cost-benefit analysis is to be truly rational, diligence must be exercised in counting previously disregarded environmental costs of production. And care must be taken to see to it that those costs are equitably distributed. Where existing markets fail, public policy may step in with a variety of instruments—effluent taxes, fines, license fees, and so forth. Environmental economists have added other important conceptual and policy innovations, for example, reopening the question of the discount rate and proposing new ways to compare the value of scarce resources and environmental quality to the present generation as compared to future ones.[11]

In its own distinctive way, then, the perspective that defines nature as a set of economic goods has made important strides in responding to the recent public outcry about the environment and quality of life. If one were to take its most ardent partisans at their words, one would have to conclude that, in fact, no other point of view has much to contribute at all. "We are going to make little real progress to solving the problem of pollution until we recognize it for what, primarily, it is: an economic problem, which must be understood in economic terms."[12] Hence, the unmistakable refrain that economists introduce in discussions about environmental values: What's it worth to you? How much are you prepared to pay for clean air and clean water? How large an investment would you like to make in preserving endangered species? If you think it's important to spend your money or public monies to save a particular wilderness area, how much are you willing to pay? What's your upper limit?

In a manner that more philosophically oriented partisans of environmentalism often find obnoxious, the economist insists that choices in this realm be understood as "trade-offs." A decision to save a coastline from offshore oil drilling is a decision to forego the possible economic payoff the oil would bring. To save the snail darter at the expense of canceling a dam construction project means giving up the economic benefits the dam would produce. Such choices are, from the point of view of economics, little more than expressions of consumer preference. Would you prefer to have condos or condors? You may not be able to afford both.[13]

Almost impossible to handle for this viewpoint are positions based on higher principles. Those who claim that there are strong moral reasons for saving this or that part of the environment are quickly advised that effective action can be extremely costly. In the words of Lester Thurow, "Environmentalism is not ethical values pitted against economic values. It is thoroughly economic. It is simply a case where a particular segment of the income distribution wants some economic goods and services (a clean environment) that cannot be achieved without collective action." For those cases in which it is difficult to determine the economic value of these environmental goods and services, Thurow suggests a "gedanken experiment," a method of "shadow pricing," to help us out. "Imagine that someone could sell you an invisible, completely comfortable facemask that would guarantee you clear air. How much would you be willing to pay for such a device?" Thus, environmental values that are not already commodities can, through this mental exercise, be seen *as if* they were commodities. In Thurow's view "the basic problem in our national debate about pollution controls is that neither side is really willing to sit down and place a value on a clean environment and then do the necessary calculations to see whether it can be had for less than this price."[14]

Hence, the advice of the economist: show me an environmental principle and I'll show you its price tag. Beyond this, economists typically have little confidence that action based upon ethical commitments will work in any case. Scoffing at the notion that "nature will be protected from man when and only when the public recognizes its moral responsibility toward nature's gifts," one economist explains the view of those in his profession: "They generally have little faith in the dependability of good intentions and pledges of virtuous behavior. Instead, economic analysis focuses on the importance of policies that work to make individual self-interest coincide with the public interest. Even when they believe that something significant can be gained by trying to change public attitudes, economists usually doubt the availability of effective means to produce such a change."[15]

In sum, the characteristic role of economists in debates about environmental issues is to call us all back to the "real world" of dollars and cents. It is all well and good, they admit, to proclaim inherent worth of wild rivers, forest areas, marshlands, endangered species, and other such things. But unless one is prepared to back up those values with real economic incentives, all is lost. Economists offer their analyses to aid us in making a choice when two or more prized environmental features are in jeopardy and perhaps not all of them can be protected or preserved. They recognize that mechanisms of public policy may be called upon as an aid in setting the money value of things. But in environmental affairs, as in business, they believe, the ultimate criterion is nothing more than the bottom line.

Nature as Endangered Ecosystem

AN ECONOMIC interpretation of nature is, of course, widely popular. Easily compatible with an urban, industrial, utilitarian, capitalist understanding of human life, this philosophy is upheld by most established institutions in modern America—corporations, governmental agencies, and even many environmental public-interest organizations as well. If different ideas of the value of nature are ever to supplant or seriously modify the prevailing system of beliefs, it will have to be because there are compelling reasons for doing so. Both of the perspectives examined next find such reasons in a view of nature inspired by the science of ecology. While the two are often grouped together as one basic philosophy, it is useful to sort out two very different varieties of argument frequently made by those who embrace ecology as a source of moral guidance.

The first of these emphasizes a theme we noted earlier in Emerson's vision: the idea that nature ultimately judges what humans do. In the late nineteenth century some Americans carried this notion to its ultimate conclusion, arguing that modern society faced a hanging judge, a grand inquisitor in an extremely vindictive mood. Thus, George Perkins Marsh's *Man and Nature* sternly proclaimed, "The ravages committed by man subvert the relations and destroy the balance which nature had established between her organic and her inorganic creations, and she avenges herself upon the intruder, by letting loose upon her defaced provinces destructive energies hitherto kept in check by organic forces destined to be his best auxiliaries, but which he has unwisely dispersed and driven from the field of action."[16] In Marsh's view it is not the mere conservation of resources that matters, not a more efficient use of things we are going to need later, but the possibility that human meddling might destroy major parts of the biosphere altogether. Twentieth-century experience has made this prospect seem plausible in a number of ways. Biologists have been able to define more precisely the complex relationships between different life forms and environmental conditions. Especially since the mid-1940s, obvious varieties of damage caused by technological applications, for instance the long-term effects of pesticides and herbicides, have been studied with scrupulous scientific care. Along with an accumulation of knowledge of this kind, our century has seen the development of weapons with unprecedented destructive power. What previous generations had feared as human encroachment upon particular natural regions, resources, and species is now overshadowed by the awful dread that we may actually eradicate all (or most) life on the planet.[17] Prophecies about the risk of ecological disaster other than nuclear holocaust carry much of the psychological force of the possible nuclear calamity. Those who portray nature as an endangered ecosystem predict, in effect, that "the Bomb" is not the only way to destroy the biosphere.[18]

The point is all too clear. If we are faced with extinction, what sense does it make to worry about nickles and dimes? Economizing on "environmental values" or weighing elaborate "trade-offs" seems foolish if catastrophe is just around the corner. As one spokesman of this point of view proclaims, "My own judgment, based on the evidence now at hand, is that the present course of environmental degradation, at least in industrialized countries, represents a challenge to the essential ecological systems that is so serious that, if continued, it will destroy the capability of the environment to support a reasonably civilized human society." At another point the same writer concludes, "Unwittingly, we have created for ourselves a new and dangerous world. We would be wise to move through it as though our lives were at stake."[19]

A number of conceivable ecological catastrophes have been described in recent years—destruction of the ozone layer, overheating of the earth through the greenhouse effect, poisoning of the oceans through a gradual buildup of industrial pollutants, and others. The likelihood or unlikelihood of each misfortune has, of course, been hotly debated. Stock in survivalist arguments rises and falls with each new shred of evidence that seems to support or deny the particular horror predicted.[20] Indeed, there is a certain vulnerability in placing the crux of one's social philosophy and policy position on the probability of eco-catastrophe. What if new data indicate the emergency wasn't what you said it was? Are you then obligated to apologize and fall silent?

It is not my task here (indeed, I simply do not have the knowledge) to evaluate the empirical validity of claims—optimistic or pessimistic—made about various eco-catastrophes. For my purposes, however, it is important to notice how the moral weight of predicted disaster provides an occasion for adopting a particular interpretation of nature—ecological theory—as a framework for understanding and judging the works of modern society. Patterns specified by the models of ecological theorists are widely taken to be "laws" that must be embraced as tenets of social wisdom. From Barry Commoner's dictum that "Nature knows best" to the detailed prescriptions of Eugene Odum about ecosystem diversity, energy flows, population growth rates, and the like, ecology leaves the realm of pure science to become a philosophy of human conduct.[21] If people would pay attention to such natural "laws" as "preserve ecosystem diversity," so those of this persuasion believe, they would have sufficient guidance to guarantee humanity's now precarious chances of survival.

At first glance it might appear that this viewpoint is nothing more than a recent version of the "naturalistic fallacy," finding the "ought" of human conduct in the "is" revealed by science. That conclusion, however, overlooks an important feature of this line of reasoning—its coercive quality. For it is not so much that we find patterns in nature that are admirable to emulate, but that ecology provides the grounds for some of our worst fears. In this way eco-philosophy has a distinctly Hobbesian twist to it. Hobbes, it is worth remembering, did not commit the "naturalistic fallacy." He simply pointed out what conditions in nature ultimately meant: a never-ending threat of violent death. On that basis he was able to argue why we should obey an absolute political authority. The *Leviathan* describes no "ought" in nature except that which arises from sheer terror. In a similar way the persuasive power of the notion of an endangered ecosystem, its claim to advise us on actions we ought to pursue, stems in the last analysis from its ability to engage our fears. If our very survival is at stake, more conventional understandings of social life must be put aside; we must rush to find effective ways of dealing with the ecological crisis.

In fact, a number of writers on ecology and society employ Hobbesian logic, coming to flagrantly Hobbesian conclusions. William Ophuls' *Ecology and the Politics of Scarcity*, for example, finds that all of us face a situation so desperate that only desperate measures will do. Traditional notions of liberty and of economic self-interest must now give way to coercive authority. "If under conditions of ecological scarcity individuals rationally pursue their material self-interest unrestrained by a common authority that upholds the common interest, the eventual result is bound to be common environmental ruin." According to Ophuls, we have a choice. Either we can establish relatively benign forms of "self coercion" right away or stumble into even more oppressive solutions later on. He recommends the creation of a class of "ecological guardians" to handle the difficult policy decisions necessary to salvage a livable world. "The individualistic basis of society," he continues, "the concept of inalienable rights, the purely self-defined pursuit of happiness, liberty as maximum freedom of action, and laissez faire itself all become problematic, requiring major modification or perhaps even abandonment if we wish to avert inexorable environmental degradation and eventual extinction as a civilization. Certainly, democracy as we know it cannot conceivably survive."[22]

It should be clear, then, how at least some prominent visions of an endangered ecosystem point to entirely different conclusions than the economic point of view. The market can no longer be relied upon as a guide; it seems to lead us straight to the "tragedy of the commons," wherein incrementally rational action eventually generates sudden doomsday. Evidently familiar social liberties must be discarded as well; they now appear reckless luxuries too risky to afford. If what the most extreme ecological survivalists say is true, there are few alternatives left other than to dismantle freedom to protect nature from assassination.

Nature as a Source of Intrinsic Good

BOTH THE environmental economist and prophet of eco-catastrophe address what they consider to be urgent environmental issues. But from another vantage point they merely scratch the surface. Many involved in today's environmental debates claim that what is needed is a radically new aesthetic, ethical, and metaphysical grasp of the human relationship to nature. They argue that unless we achieve this new vision of our situation, we will surely fail to make the fundamental changes—changes in the very foundations of our culture—that true ecological wisdom demands. Drawing upon philosophies of nature and religious ideas of earlier times, such as those of Tao, Zen, St. Francis, St. Benedict, Henry David Thoreau, John Muir, and Alfred North Whitehead among others, a group of contemporary eco-philosophers have sought a renewed reverence for things natural and ways of "following nature" for positive moral instruction.[23]

A milestone in the popular history of this way of thinking was the publication of Lynn White's "The Historical Roots of Our Ecological Crisis" in *Science* in 1967.[24] Widely reprinted and widely discussed ever since, the article made an eloquent case for the view that present-day problems of pollution, resource depletion, environmental degradation, and the like reflect our culture's most fundamental stance toward the material world, a stance deeply rooted in Western metaphysics. Many have disagreed with White's analysis of the Christian origin of the malady; many have rejected the specifics of his solution—a reaffirmation of the teachings of St. Francis—as a flimsy nostrum. But the article did provide an invitation to a style of inquiry that a great many have taken to heart.

By the early 1970s it was possible for the Norwegian philosopher Arne Naess to proclaim that there are essentially two approaches to environmental issues. What he called "shallow environmentalism" tries to remedy the most obvious damage done by urban industrial society. This approach is directed toward reducing pollution and husbanding resources more intelligently than was done in the recent past. Its central goal, in Naess' view, is the continuing affluence of the industrialized nations. In contrast, he identifies "deep ecology." It too is concerned with reducing pollution and wasting of resources, but also seeks an orientation in ethics, politics, and culture that rejects the destructiveness of the urban/industrial way of life to pursue a more positive relationship to the biosphere.[25]

A number of attempts have been made to satisfy the desire for a vision of "deep ecology," theories that range from poetic speculation to tightly reasoned arguments in analytical philosophy. Common to this enterprise, however, and absolutely crucial to its central claims is a defense of biological egalitarianism. Anthropocentric beliefs of "man in nature" and "nature for man" are rejected as pure ignorance; human beings stand as merely one species among millions. There is no good reason to suppose that our species has any special right to rule the rest of creation; nothing but sheer hubris supports that prejudice. The Baconian presumption that humans are rightful conquerors and masters of nature and the Lockean argument that humans have legitimate "property" in the wild are both ideas that must be rejected as perverse overestimates of our worldly status. Even the hopeful view advocated by René Dubos and others that man's best role would be stewardship over natural creation, a role that finds us actually improving the earth by wise cultivation and engineering, is one that many eco-philosophers find distasteful; "stewardship" represents, they claim, anthropocentric dominance smuggled back into the picture under a more attractive label.[26] Needed instead, they insist, is a truly ecological ethic, one that sees human beings as partners with natural entities—animate and inanimate—all of roughly equal standing.

The conception of nature usually favored in this context is, again, that of the "ecosystem" developed in field ecology. But unlike those who take this model primarily as a guide to survival, proponents of deep ecology interpret it as a general framework of social norms carrying the imprimatur of science. Generally recognized as spiritual father of this orientation is game manager and ecologist Aldo Leopold. In the 1940s Leopold announced his reasons for abandoning standard notions of resource conservation. "A system of conservation based solely on economic self-interest is hopelessly lopsided. It tends to ignore, and thus eventually to eliminate, many elements in the land-community that lack commercial value, but that are (as far as we know) essential to its healthy functioning." As an alternative he proposed an "ecological conscience" and "land ethic" that would change "the role of *Homo sapiens* from conqueror of the land-community to plain member and citizen of it."[27]

Not content with vague musings about harmony or unity with nature, Leopold tried to define ethical imperatives that stem from an acknowledgment that humans are parts of a complex ecological world of interdependent parts. Foremost among the maxims he formulated is one that specifies, "A thing is right when it tends to preserve the integrity, stability, and beauty of the biotic community. It is wrong when it tends otherwise."[28]

Following this lead, a number of philosophers and scientists have tried to translate the terms of field ecology into a coherent moral philosophy. Undaunted by the possibility that naturalistic fallacies might wreck the quest, many have taken such concepts as entropy, homeostasis, and ecological diversity from scientific disciplines and used them to justify ethical prescriptions. This is not to say that such inquiries have been done in an uncritical way. The enormous literature on environmental ethics contains equally enormous disagreements about which reasons are valid and which are not. For every serious-minded thinker who believes to have discovered the much-desired "new ethic," there is an equally serious-minded person who denies it. Although eco-philosophers are sometimes criticized as true believers, their discussions contain much less ideological uniformity than, for example, economic approaches to environmental issues.

From the standpoint of deep ecology the values of modern society are not so much wrong as they are disoriented. We no longer understand which kinds of natural entities ought to be, in principle, removed from the sphere of economic transactions altogether. We fail to see, for example, that the market value of a mountain filled with minerals is not truly comparable to the intrinsic value that mountain has as part of a wilderness region. Even the distinction between "public interest" as opposed to "private interest" as defined by liberal political economy is of little help in acknowledging and preserving the intrinsic value of natural things. Rigorous arguments to the effect that "animals have rights" and "trees have standing" may begin to restore the ability to achieve a balanced judgment.[29] But for a society that has gotten used to all things as potential commodities to be mined, developed, processed, packaged, marketed, used, and discarded, clearly more would be needed to turn things around than just a new set of clever arguments.

That is why the ideas of deep ecology, whatever their philosophical merit may be, are basically appeals to the heart. "I believe in wilderness for itself alone," one leading spokesman of this perspective exclaims. "I believe in the rights of creatures other than man."[30] Cares of this kind are expressed in practice in many different ways. Some are content to remain within the framework of existing environmental interest groups, the Sierra Club and Friends of the Earth, for example, working to achieve the more idealistic preservationist goals of those organizations. Others find established channels—election campaigns, legislative lobbying, government policy making, and the like—largely irrelevant. Of these, some devote their efforts to investigations in social theory, utopian fiction, or scientific research that try to describe what a society based upon ecologically sound principles would actually be. Others have gone about forming new communities or re-forming old ones in ways that manifest their favored version of this hope.

But whatever the specific path chosen, advocates of deep ecology tend to believe that nothing less than the strongest commitment will make any difference at all. If we have no feeling for natural creations outside their instrumental value, there can be little hope for sweeping changes in our culture's basic environmental posture. And if our concerns here are based only in fears for our own survival, then the changes we might make are less likely to be beautiful than hideous.

Nature as a Social Category

THE IDEAS I have outlined by no means encompass all the positions to be found in contemporary environmentalism; they are best understood as points along a broad spectrum of thought. We have seen enough, however, to understand that serious differences in basic orientation exist in a movement often portrayed as a unified, single-minded whole. My point is not to identify any particular position or group or spokesman as having the greater share of wisdom. Indeed, all three tendencies of mind are often present simultaneously, their conflicts unresolved, in the thinking of individuals and groups who call themselves environmentalists or ecologists.

In their most purely ideological moments each of the three dispositions finds much to scorn in one or more of the others. Environmental economics, unmodified by any broader sentiments, has little patience with attempts to defend unquantified, unmonetized, aesthetic values in nature. It tends to be suspicious of the motives of persons who find intrinsic value in the wild. "Being comfortable but not rich," one particularly cynical book complains, "environmentalists may welcome using other people's money to keep competitors away from the wilderness."[31] In turn, many committed ecological survivalists and deep ecologists express their contempt for those who have not embraced what they uphold as the true path. Hence, one prominent eco-philosopher has argued that mere reformist environmentalism is a manifestation of the problem, not its cure. "Environmentalism does not question the most basic premise of the present society, notably, that humanity must dominate nature; rather, it seeks to facilitate that notion by developing techniques for diminishing the hazards caused by the reckless despoilation of the environment." [32]

But when the charges and countercharges subside, one often finds different sides in the debate borrowing each other's currency. It is common to find that an environmental policy study filled with straightforward utilitarian analyses of environmental topics neverthless begins with a rhapsodic preface that sounds as if it could have been written by Aldo Leopold or John Muir. Similarly, those who hold more radical environmental views often bolster their case with economic arguments, taking pains to describe ways in which endangered plant and animal species might have great practical use. Two writers well known for their strong ecological survivalism try to bolster their case with cash register refrains, arguing that "populations of organisms and entire species are heedlessly being exterminated around the world at an increasing rate . . . with no regard whatever for their potential source of food, fiber, drug, or any of a myriad other useful substances, such as spices, oils, industrial chemicals, hides, and so on)." [33] For a society responsive to commercial values, such arguments may make sense. At the same time, however, they create a disingenuous tone in environmental advocacy. It is as if those who had come to worship at the temple had decided to change a little money on the side.

"Nature," Georg Lukacs once observed, "is a societal category." [34] That judgment overstates the case, but not by very much. It is certainly true that a material reality with its own inherent properties does exist beyond the human power to investigate, interpret, and manipulate that reality. Nothing the human mind or social institutions accomplish can alter the simple presence of "nature" as a raw "totality of things." Beyond that brute fact, however, there are enormous possibilities for interpreting nature and society as reciprocal realms of meaning. Again and again one finds that metaphors first used to describe the one are later translated to illuminate the other. In our time discussion about ecology and environment tell us a great deal more about the condition of society than about anything in nature as such. The ecological persuasion, in both its survivalist and deep ecology versions, has challenged an idea long prevalent in Western culture, an idea that portrays nature as an object of control and source of wealth. The alternative model now proposed holds that nature be seen as a system, an incalculably intricate, delicately balanced aggregate of interdependent parts and processes. Offering this idea as a true picture of the world, the ecologist may tend to forget that the model is itself a human creation, an abstract representation of certain kinds of phenomena. This creation itself reflects a great deal about today's man–made world; in some respects, the ecological model presents a mirror image of advanced industrial society. An ecosystem is a utopia of sorts, doing well what artificial structures do poorly. Thus, the ecosystem manages its complexity in ways that guarantee optimal results; unlike human organizations, there are no wrong-headed decisions to foul things up. Ecosystem change has benign homeostasis as its telos; by contrast, change in modern society produces chaotic disruptions that never cease. The self-governing processes of any ecosystem are a wonder to behold; twentieth-century technological societies often appear beyond governance altogether.

As we have seen, this ecological utopia has been employed to justify two distinctly different kinds of social advice. One argument offers the counsel of caution born of fear. What we had thought was a mere border skirmish with nature turns out to be something much more serious, a battle whose outcome is likely to be the destruction of both sides. Faced with the prospect of our own extinction, we humans must now come to our senses and find ways to disarm the powerful weapons poised at the heart of the ecosystem. According to some observers, this will require establishing overwhelmingly strong state control. Another school of thought, however, uses essentially the same conception of nature to offer a counsel of love. In its eyes what is significant about ecosystems is simply that they are good. If we learned to comprehend, respect, and cherish this beneficence, we would discover a fully adequate guide for living. From this vantage point it is not merely a detente in our struggle against nature that is needed, but the widespread adoption of harmonious social patterns that the study of ecology appears to sanction.

The reflection of social conditions, issues, policies, and utopias in ideas of the natural is something more easily recognized in earlier periods than in our own. "Nature, red in tooth and claw" where only the fittest survive now seems to us a bizarre distortion from the era of nineteenth-century imperialism and robber baron capitalism. But there may come a time in which our own attempts to live according to ecological principles will seem equally misguided. Today, as in the past, ideas about things natural must be examined and criticized not only for ways they help us understand the material world, but for the quality of their social and political counsel. Nature will justify anything. Its text contains opportunities for myriad interpretations. The patterns noticed in natural phenomena and the meanings given them are all matters of choice. We must learn to read contemporary interpretations of the environment and ecology as we read Hobbes, Locke, or Rousseau on "the state of nature," to see exactly what notion of society is being chosen. When that is done, nature and social forms can be evaluated separately, a practice that an awareness of many past mistakes strongly recommends. It is comforting to assume that nature has somehow been enlisted on our side. But we are not entitled to that assumption.

8

ON NOT HITTING
THE TAR-BABY

Every time you make a move, you take a risk.

Monty Hall
(host, "Let's Make a Deal")

THE MOST PREVALENT way our society explores the possibility of limiting technology is through the study of "risk." Noting how the broader effects of industrial production can damage environmental quality and endanger public health and safety, risk assessment seeks to perfect methods of evaluation that are at once rigorous and morally sound. This approach appears to offer policy makers a way to act upon the best scientific information to protect society from harm. Indeed, if we define "risk" as everything that could conceivably go wrong with the use of science and technology—a definition that many are evidently prepared to accept—then it seems possible that we might arrive at a general understanding of norms to guide the moral aspects of scientific and technical practice.

But the promise of risk assessment is difficult to realize. The arena in which discussions of risk take place is highly politicized and contentious. Specific questions such as those dealing with the safety of nuclear power, as well as more general ones having to do with choosing proper methodologies for studying risks at all, involve high stakes. Powerful social and economic interests are invested in attempts to answer the question, How safe is safe enough? Expert witnesses on different sides of such issues are often best identified not by what they know, but rather whom they represent. Indeed, the very introduction of "risk" as a common way of defining policy issues is itself far from a neutral issue. At a time in which modern societies are beginning to respond to a wide range of complaints about possible damage various industrial practices have on the environment and public health, the introduction of self-conscious risk assessment adds a distinctly conservative influence. By the term "conservative" here I mean simply a point of view that tends to favor the status quo. Although many of those who have become involved in risk assessment are not conservative in a political sense, it seems to me that the ultimate consequence of this new approach will be to delay, complicate, and befuddle issues in a way that will sustain an industrial status quo relatively free of socially enforced limits. It is the character of this conservatism that I want to explore here.

Hazards and Consensus

AS COMPARED TO other varieties of moral and political argument, risk assessment seeks a very narrow consensus. It asks us to evaluate circumstances in which there is some chance, perhaps a very remote chance, of harm from activities that are assumed to be socially beneficial in other respects. If one is able to recognize and care about the possibility of such harm, one is eligible to enter the discussion. No other grounds of agreement are necessary. It does not matter what one's views on other issues may be, what one thinks about arms control, deficit spending, or abortion, for instance. People who have drastically different views on, say, welfare payments for the poor and virtually every other social question may nevertheless find reason to act together on shared dangers to health and safety.

In this respect risk assessment does not seek to provide a general evaluation of the conditions of modern life, as did, for example, liberalism, Marxism, or other broad-scale social theories. It sometimes happens, of course, that the question of risk is formulated as a problem to be discussed within the categories of a more comprehensive theory. A Marxist may reformulate the problem as a wrinkle in the analysis of relations of production; a utilitarian may wish to see it as a source of perplexity in the goal to achieve the greatest good for the greatest number. But such discussions are not prominent in claims and counterclaims about specific hazards. We do not look to debates about DDT, PCBs, air pollution, nuclear power, and the like for a comprehensive grasp of the modern condition. The topic here concerns, as one observer has described it, "making industrialism safe for human life."[1]

But although the consensus that risk assessment seeks is a narrow one, it is potentially very strong. The fundamental issue here is fear—fear of injury, disease, death, and the prospect of having to live in deteriorating surroundings. For that reason arguments about impending dangers are often useful in attempts to unite people who have little in common other than shared fears. Contemporary environmentalism and consumerism have taken advantage of this opportunity to marshal support for their causes. On a much different scale the military has won legitimacy for its multibillion dollar weapons projects by engaging the public's anxieties about ambiguously defined enemies and the peril of nuclear war.

As noted in the last chapter, for stark appreciation of arguments predicated on fear, there is no better counsel than the great political psychologist of dread, Thomas Hobbes. It was Hobbes' fundamental insight that people who can agree on nothing else will neverthless recognize that they share a morbid fear of physical harm from each other. Even the strongest person in the state of nature is vulnerable to attack in an unguarded moment. When people acknowledge the continuing terror that surrounds them, they will be receptive to entering a compact to establish political society and its reasonable system of authority and obligation. As the first and most brilliant modern writer on risk assessment, Hobbes gives us adequate reasons why people's fears should never be taken lightly.

Numerous examples in the history of modern industrial society lend support to the belief that overtly dangerous applications of new technology will not be long tolerated. Head-on collisions of passenger trains in New England in the mid–nineteenth century called attention to scheduling and communications problems for railroads of the time. The trains in question ran on single lines of track, a situation that made simultaneous two-way traffic a tricky matter even with occasional switchovers. Regulations passed by state legislatures enforced a widely appreciated need for remedies.[2] A similar set of events afflicted the early history of jet airline travel. An unforeseen weakness in the aluminum body of the British *Comet* caused several of the planes to fall from the sky. Study of the wreckage revealed the problem's source, and the manufacturer redesigned its planes in response. Thus, the early stages of development of new machines, chemicals, techniques, and large-scale systems frequently involve a period of trial and error in which people are killed or injured. A commonly accepted norm is that obvious sources of harm must be eliminated through either private or public action; otherwise the very usefulness of the device will be called into question. And there are examples of technology—thalidomide is an obvious case—ultimately judged not a tool, but a menace.

In fact, the public reponse to dangerous effects of modern technological and industrial systems has often been a stimulus for reform when other sources were weak. Upton Sinclair's *The Jungle* (1906) aroused widespread protest about unhealthy conditions in the Chicago stockyards. Investigations by public officials led to new laws and remedies. Sinclair, a socialist, wrote dozens of books decrying the evil effects of capitalism in modern society. None of his later writings, however, was as successful in capturing the popular imagination or in stimulating change than his exposé of the meat-packing industry.[3] From Sinclair's experience and that of other more recent political activists, it is clear that alarms about particular hazards will engage the public's imagination where more ambitious, general criticisms do not. Hence, the politics of hazards often becomes a strategic complement for or even an alternative to the politics of social justice.

The attempt to parlay the principle *salus populi suprema lex* into a full-blown political movement has become a familiar approach among contemporary activists. By calling attention to a possible danger, one hopes to attract support for a broader program of social criticism and reform. The alleged danger works as a symbol that may enable people to consider other social maladies, for example, the concentration of institutional power. Thus, the first sentence of Ivan Illich's *Medical Nemesis* proclaims, "The medical establishment has become a major threat to health."[4] Illich's method is to focus on iatrogenic (physician caused) disease as a way to attract the reader's attention to his primary concern: the destructive social organization of modern medicine. Clearly, in Illich's view, this mode of organization—monolithic, bureaucratic, expert centered—is pernicious wherever it occurs. The fact that it is also the source of certain identifiable health problems gives a compelling reason to investigate the problem. Illich seizes upon this fact to instruct his audience about the oppressive social structures of modern life, not just in medicine, but throughout a whole range of institutional settings.

Similar approaches are used with varying success by environmentalists, consumerists, and muckraking political journalists. Headline news about dangers to health and safety lend an opportunity to discuss what these observers describe as even more fundamental issues, for example, the need to control the enormous power of large business corporations. Articles and editorials frequently published in *The Nation, The Progressive, Mother Jones,* or *In These Times* show this persuasive strategy at work. At one level it is perfectly sound. In a society strongly committed to capitalism as a way of life it is difficult to address the social ills of capitalist practices head on. A more subtle tack is to begin discussing urgent issues that do not appear to have any ideological valence at all.[5] In pointing to this strategy I do not wish to dismiss any of the claims made by these political activists or writers. But it is fair to say that this approach, much like the "efficiency" arguments that radicals sometimes make, is another example of a conceptual Trojan horse. If it were effective to tackle social injustice and the concentration of economic power directly, I suppose we would do it more often.

Risk and Fortitude

BUT AS political strategies sometimes do, this way of addressing social issues can backfire. The rise of risk assessment in the 1970s gave adequate notice that a strong backfire had begun. Questions that had previously been talked about in such terms as the "environmental crisis," "dangerous side effects," "health hazards," and the like were gradually redefined as questions of "risk."[6] The difference is of no small importance.

If we declare ourselves to be identifying, studying, and remedying hazards, our orientation to the problem is clear. Two assumptions, in particular, appear beyond serious question. First, we can assume that given adequate evidence, the hazards to health and safety are fairly easily demonstrated. Second, when hazards of this kind are revealed, all reasonable people usually can readily agree on what to do about them. Thus, if we notice that a deep, open pit stands along a path where children walk to school, it seems wise to insist that the responsible party, be it a private person or public agency, either fill the pit or put a fence around it. Similarly, if we have good reason to believe that an industrial polluter is endangering our health or harming the quality of the land, air, or water around us, it seems reasonable to insist that the pollution cease or be strongly curtailed. Straightforward notions of this kind, it seems to me, lie at the base of a good many social movements concerned with environmental issues, consumer protection, and the control of modern technology. In their own ways, of course, such movements are capable of adding elements of complication to policy discussion, for example, notions of complexity from ecological theory. Typically, however, these complications are ones that ultimately reinforce a basic viewpoint that sees "dangers" to human health, other species, and the environment as grave matters that are fairly easy to understand and require urgent remedies.

If, on the other hand, we declare that we are interested in assessing risks, complications of a different sort immediately enter in. Our task now becomes that of studying, weighing, comparing, and judging circumstances about which no simple consensus is available. Both of the commonsense assumptions upon which the concern for "hazards" and "dangers" rely are abruptly suspended. Confidence in how much we know and what ought to be done about it vanishes in favor of an excruciatingly detailed inquiry with dozens (if not hundreds) of fascinating dimensions. A new set of challenges presents itself to the scientific and philosophical intellect. Action tends to be postponed indefinitely.

As one shifts the conception of an issue from that of hazard/danger/threat to that of "risk," a number of changes tend to occur in the way one treats that issue. What otherwise might be seen as a fairly obvious link between cause and effect, for example, air pollution and cancer, now becomes something fraught with uncertainty. What is the relative size of that "risk," the "chance of harm"? And what is the magnitude of the harm when it does take place? What methods are suited to measuring and analyzing these matters in a suitably rigorous way? Because these are questions that involve scientific knowledge and its present limits, the risk assessor is constrained to acknowledge what are often highly uncertain findings of the best available research. For example, one must say in all honesty, "We don't know the relationship between this chemical and the harm it may possibly cause." Thus, the norms that regulate the acceptance or rejection of the findings of scientific research become, in effect, moral norms governing judgments about harm and responsibility. A very high premium is placed on not being wrong. Evidence that "the experts disagree" adds further perplexity and a need to be careful before drawing conclusions. The need to distinguish "facts" from "values" takes on paramount importance. Faced with uncertainty about what is known concerning a particular risk, prudence becomes not a matter of acting effectively to remedy a suspected source of injury, but of waiting for better research findings.[7]

An illustration of this cast of mind can be seen in a study done for the Environmental Protection Agency to determine whether or not there were indications that residents of the Love Canal area of New York, an abandoned chemical waste disposal site, showed chromosome damage. The report written by Dante Picciano, a geneticist employed by the Biogenics Corporation of Houston, Texas, drew the following conclusions: "It appears that the chemical exposures at Love Canal may be responsible for much of the apparent increase in the observed cytogenetic aberrations and that the residents are at an increased risk of neoplastic disease, of having spontaneous abortions and of having children with birth defects. However, in absence of a contemporary control population, prudence must be exerted in the interpretation of such results."[8] Although the chemicals themselves may have been disposed of in reckless fashion, scientific studies on the consequences must be done with scrupulous care. Insofar as law and public policy heed the existing state of scientific knowledge about particular risks, the same variety of caution appears in those domains as well.

Frequently augmenting these uncertainties about cause and effect are the risk assessor's calculations on costs and benefits. To seek practical remedies for man-made risks to health, safety, or environmental quality typically requires an expenditure of public or private money. How much is it reasonable to spend in order to reduce a particular risk? Is the cost warranted as compared to the benefit received? Even if one is able to set aside troubling issues about equity and "who pays," risk/cost/benefit calculations offer, by their very nature, additional reasons for being hesitant about proposing practical remedies at all. Because it's going to cost us, we must ponder the matter as a budget item. Our budgets, of course, include a wide range of expenditures for things we need, desire, or simply cannot avoid. Informed about how the cost of reducing environmental risks is likely to affect consumer prices, taxes, industrial productivity, and the like, the desire to act decisively with respect to any particular risk has to be weighed against other economic priorities.[9]

A willingness to balance relative costs and benefits is inherent in the very adoption of the concept of "risk" to describe one's situation. In ordinary use the word implies "chance of harm" from the standpoint of one who has weighed that harm against possible gain. What does one do with a risk? Sometimes one decides to *take it*. What, by comparison, does one do with a hazard? Usually one seeks to avoid it or eliminate it. The use of the concept of "risk" in business dealings, sports, and gambling reveals how closely it is linked to the sense of voluntary undertakings. An investor risks his capital in the hope of making a financial gain. A football team in a close game takes a risk when it decides to run on fourth down and a yard to go. A gambler at a Las Vegas blackjack table risks his or her money on the chance of a big payoff. In contrast to the concepts of "danger," "hazard," or "peril," the notion of "risk" tends to imply that the chance of harm in question is accepted willingly in the expectation of gain. This connotation makes the distinction between voluntary and involuntary risks outlined in some of the recent literature largely misleading. The word carries a certain baggage, a set of ready associations. The most important of these is the simple recognition that all of us take risks of one kind or another rather frequently.

Noticing that everyday life is filled with risky situations of various kinds, contemporary risk assessment has focused upon a set of psychological complications that further compound the difficulties offered by scientific uncertainty and the calculations of risk/cost/benefit analysis. Do people accurately assess the risks they actually face? How well are they able to compare and evaluate such risks? And why do they decide to focus upon some risks rather than others? A good deal of interesting and valid psychological research has been devoted to answering such questions. By and large, these studies tend to show that people have a fairly fuzzy comprehension of the relative chance of harm involved in their everday activities.[10] If one adds to such findings the statistical comparisons of injuries and fatalities suffered in different situations in modern life, then the question of why people become worried about certain kinds of risks and not others becomes genuinely puzzling.[11]

The rhetorical possibilities of this puzzle are often seized upon by writers who assert that people's confusion about risks discredits the claims of those who focus upon the chance of harm from some particular source. Why should a person who drives an automobile, a notorious cause of injury and death, be worried about nuclear power or the level of air pollution? Invidious comparisons of this kind are sometimes employed to show that people's fears about technological hazards are completely irrational. Hence, one leading proponent of this view argues, "it is not surprising that people with psychological and social problems are unsettled by technological advance. The fears range from the dread of elevators in tall buildings to apprehension about 'radiation' from smoke detectors. Invariably, these fears are evidence of displacing of inner anxiety that psychiatrists label as phobic." The same writer explains that normal folk are able to overcome such phobias by reminding themselves of the incalculable good that modern technologies have brought to all of us. "People of sound mind accept the negligible risk and minor inconvenience that often go hand in hand with wondrous material benefits."[12]

Once one has concluded that reports about technological risks are phobia based, the interesting task becomes that of explaining why people have such fears at all. Tackling this intellectual challenge, anthropologist Mary Douglas and political scientist Aaron Wildavsky have developed a style of analysis based on the assumption that complaints about risk are not to be taken at face value. In their view all reports about environmental risks must be carefully interpreted to reveal the underlying social norms and institutional attachments of those making the complaints. Different kinds of institutions respond to risk in very different ways. For example, entrepreneurs accept many kinds of economic risk without question. They embrace the invigorating uncertainties of the market, the institutional context that gives their activities meaning. In contrast, public-interest organizations of the environmental movement, organizations that Douglas and Wildavsky describe as "sects," show, in their view, obsessive anxiety about technological risks; the discovery of these risks provides a source of personal commitment and social solidarity the "sects" so desperately need. Are there any environmental dangers in the world that all reasonable people, regardless of institutional attachment, ought to take seriously? Douglas and Wildavsky find that question impossible to answer. The fact that "the scientists disagree" requires us to be ever skeptical about any claims about particular risks. Instead, Douglas and Wildavsky offer the consolations of social scientific methodology to help us explain (and feel superior to) the strange behavior of our benighted contemporaries.[13]

Entering thickets of scientific uncertainty, wending our way through labyrinths of risk/cost/benefit analysis, balancing skillfully along the fact/value gap, stopping to gaze upon the colorful befuddlement of mass psychology, we finally arrive at an unhappy destination—the realm of invidious comparison and social scorn. This drift in some scholarly writings on risk assessment finds its complement in the public statements and advocacy advertising of corporations in the oil, chemical, and electric power industries. In the late 1970s the debunking of claims about environmental hazards became a major part of corporate ideology. Closely connected to demands for deregulation and the relaxing of government measures to control air pollution, occupational safety and health, and the like, the "risk" theme in the pronouncements of industrial firms assumed major importance.

A typical advertisement from Mobil Oil's "Observations" series illustrates the way in which popularized risk psychology and risk/cost/benefit analysis can work in harmony. "Risky business," the ad announces. "Lawn mowers . . . vacuum cleaners . . . bathtubs . . . stairs . . . all part of everyday life and all hazardous to your health. The Consumer Product Safety Commission says these household necessities caused almost a million accidents last year, yet most people accept the potential risks because of the proven benefits. . . . Risk, in other words, is part of life. Fool's goal. Nothing's safe all the time, yet there are still calls for a 'risk-free society.'" Although I have read large portions of the recent literature on energy, environment, consumer protection, and the like, I cannot recall having seen even one instance of a demand for a "risk-free" society. The notion appears only as a straw man in advocacy ads like this one. Its text goes on to evoke a string of psychological associations linked to the experience of "risk" in economic enterprise. "Cold feet. What America does need are more companies willing to take business risks, especially on energy, where the risks are high. . . . We're gamblers. . . . Taking risks: it's the best way to keep America rolling . . . and growing."[14] Poker anyone?

There is, then, a deep-seated tendency in our culture to appreciate risk-taking in economic activity as a badge of courage. Putting one's money, skill, and reputation on the line in a new venture identifies that person as someone of high moral character. On the other hand, people who have qualms about the occasional side effects of economic wheeling and dealing can easily be portrayed as cowardly and weak-spirited, namby-pambies just not up to the rigors of the marketplace. Public policies that recognize such qualms can be dismissed as signs that the society lacks fortitude or that the citizenry has grown decadent. Thus, in addition to other difficulties that await those who try to introduce "risk" as a topic for serious political discussion, there is a strong willingness in our culture to embrace risk-taking as one of the warrior virtues. Those who do not possess this virtue should, it would seem, please not stand in the way of those who do.

Avoiding "Risk"

BY CALLING ATTENTION to these features in contemporary discussions about risk, I do not want to suggest, as some have done, that the whole field of study has somehow been corrupted by the influence of selfish economic interests.[15] Neither am I arguing that all or most conversations on this topic show a deliberate, regressive political intent. Indeed, many participants have entered the debate with the most noble of scientific, philosophical, and social goals. Much of the analytically solid writing now produced on this topic seeks to strengthen intellectual armaments used to defend those parts of society and the environment most likely to experience harm from a variety of technological side effects.[16] And certainly there are many fascinating issues under the rubric of risk assessment that are well worth pursuing. I can only join in wishing that such clearheaded, magnanimous work flourish.

But from the point of view I've described here, the risk debate is one that certain kinds of social interests can expect to lose by the very act of entering. In our times, under most circumstances in which the matter is likely to come up, deliberations about risk are bound to have a strongly conservative drift. The conservatism to which I refer is one that upholds the status quo of production and consumption in our industrial, market-oriented society, a status quo supported by a long history of economic development in which countless new technological applications were introduced with scant regard to the possibility that they might cause harm. Thus, decades of haphazard use of industrial chemicals provide a background of expectations for today's deliberations on the safety of such chemicals. Pollution of the air, land, and water are not the exception in much of twentieth-century America, but rather the norm. Because industrial practices acceptable in the past have become yardsticks for thinking about what will be acceptable now and in the future, attempts to achieve a cleaner, healthier environment face an uphill battle. The burden of proof rests upon those who seek to change long-existing patterns.

In this context, to define the subject of one's concerns as a "risk" rather than select some other issue skews the subsequent discussion in a particular direction. This choice makes it relatively easy to defend practices associated with high levels of industrial production; at the same time it makes it much more difficult for those who would like to place moral or political limits upon that production to make much headway. I am not saying that this is a consequence of the way risk assessment is "used," although conservative uses of this sort of analysis are, as we have seen, easily enough concocted. What is more important to recognize is that in a society like ours discussions centering on risk have an inherent tendency to shape the texture of such inquiries and their outcome as well. The root of this tendency lies, very simply, in the way the concept of "risk" is employed in everyday language. As I have noted, employing this word to talk about any situation declares our willingness to compare expected gain with possible harm. We generally do not define a practice as a risk unless there is an anticipated advantage somehow associated with that practice. In contrast, this disposition to weigh and compare is not invoked by concepts that might be employed as alternatives to "risk"—"danger," "peril," "hazard," and "threat." Such terms do not presuppose that the source of possible injury is also a source of benefits. From the outset, then, those who might wish to propose limits upon any particular industrial or technological application are placed at a disadvantage by selecting "risk" as the focus of their concerns. As they adopt risk assessment as a legitimate activity, they tacitly accept assumptions they might otherwise wish to deny (or at least puzzle over): that the object or practice that worries them must be judged in light of some good it brings and that they themselves are recipients of at least some portion of this good.

Once the basic stance and disposition associated with "risk" have defined the field of discourse, all the complications and invidious comparisons I have described begin to enter in. Standards of scientific certainty are applied to the available data to show how little we know about the relationship of cause and effect as regards particular industrial practices and their broader consequences. Methods of risk/cost/benefit analysis fill out a detailed economic balance sheet useful in deciding how much risk is "acceptable." Statistical analyses show the comparative probability of various kinds of unfortunate events, for example, being injured in a skiing accident as compared to being injured by a nuclear power plant meltdown. Psychological studies reveal peculiarities in the ways people estimate and compare various kinds of risks. Models from social science instruct us about the relationship of institutional structures to particular objects of fear. A vast, intricately specialized division of intellectual labor spreads itself before us.

One path through this mass of issues is to take each one separately, seeking to determine which standards, methods, findings, and models are appropriate to making sound judgments about problems that involve public health, safety, and environmental quality. For example, one might question how reasonable it is to apply the very strict standards of certainty used in scientific research to questions that have a strong social or moral component. Must our judgments on possible harms and the origins of those harms have only a five percent chance of being wrong? Doesn't the use of that significance level mean that possibly dangerous practices are "innocent until proven guilty"?[17]

Similarly, one might reevaluate the role that cost/benefit analysis plays in the assessment of risks, pointing to the strengths and shortcomings of that method. How well are we able to measure the mix of "costs" and "benefits" involved in a given choice? What shall we do when faced with the inadequacy of our measurements? Are criteria of efficiency derived from economic theories sufficient to guide value choices in public policy? In controversies about the status of the intellectual tools used in decision making, such questions are hotly disputed.[18]

But for those who see issues of public health, safety, and environmental quality as fairly straightforward matters requiring urgent action, these exercises in methodological refinement are of dubious value. It is sensible to ask, Why get stuck in such perplexities at all? Should we spend our time working to improve techniques of risk analysis and risk assessment? Or should we spend the same time working more directly to find better ways to secure a beautiful, healthy, well-provided world and to eliminate the spread of harmful residues of industrial life?

The experience of environmentalists and consumer advocates

who enter the risk debate will resemble that of a greenhorn who visits Las Vegas and is enticed into a poker game in which the cards are stacked against him. Such players will be asked to wager things very precious to them with little prospect that the gamble will deliver favorable returns. To learn that the stacked deck comes as happenstance rather than by conscious design provides little solace; neither will it be especially comforting to discover that hard work and ingenuity might improve the odds somewhat. For some, it is simply not the right game to enter. There are some players at the table, however, who stand a much better chance. Proponents of relaxed government regulations on nuclear power, industrial pollution, occupational safety and health, environmental protection, and the like will find risk assessment, insofar as they are able to interest others in it, a very fruitful contest. Hence, Chauncey Starr, engineer and advocate of nuclear power, is well advised to take "risk" as the central theme in his repertoire of argument. But the likes of David Brower, Ralph Nader, and other advocates of consumer and environmental interests would do well to think twice before allowing the concept to play an important role in their positions on public issues.

Fortunately, many issues talked about as risks can be legitimately described in other ways. Confronted with any cases of past, present, or obvious future harm, it is possible to discuss that harm directly without pretending that you are playing craps. A toxic waste disposal site placed in your neighborhood need not be defined as a risk; it might appropriately be defined as a problem of toxic waste. Air polluted by automobiles and industrial smokestacks need not be defined as a "risk"; it might still be called by the old-fashioned name, "pollution." New Englanders who find acid rain falling on them are under no obligation to begin analyzing the "risks of acid rain"; they might retain some Yankee stubbornness and confound the experts by talking about "that destructive acid rain" and what's to be done about it. A treasured natural environment endangered by industrial activity need not be regarded as something at "risk"; one might regard it more positively as an entity that ought to be preserved in its own right.

About all these matters there are rich, detailed forms of discourse that can strengthen our judgments and provide structure for public decisions. A range of theoretical perspectives on environmental protection, public health, and social justice can be

151

drawn upon to clarify the choices that matter. My suggestion is that before "risk" is selected as a focus in any area of policy discussion, other available ways of defining the question be thoroughly investigated. For example, are health and safety hazards that blue-collar workers encounter on the job properly seen as a matter of "risk" to be analyzed independently of directly related economic and social conditions? Or is it more accurate to consider ways in which these hazards reflect a more general set of social relationships and inequalities characteristic of the free enterprise system? One's initial definition of the problem helps shape subsequent inquiries into its features. If one identified the issue of worker health and safety as a question of social justice, there would be less need to do all of the weighing of probabilities, comparing of individual psychological responses, and performing of other delicate tasks that risk assessment involves. It might still be interesting to do research on levels of air pollution in executive offices as compared to those in factories. Of course, one always wants to have the best scientific information on such issues. But, in all likelihood, such studies would reveal little new or surprising. It is common knowledge that our society distributes wealth, income, knowledge, and social opportunities unequally. To establish that it also distributes workplace hazards inequitably merely amplifies the problem. Those concerned with questions of social justice would do well to stick to those questions and not look to risk analysis to shed much light.

My point is, then, that a number of important social and political issues are badly misdefined by identifying them as matters of "risk." Whenever possible, such misdefinitions ought to be resisted along with the methodological quagmires they entail. This is not to say that there are no issues of broad-scale social policy in which the concept of "risk" is legitimately applied. There are some applications of modern science and technology in which the uncertainty that surrounds suspect practices and their possible effects is so great that "risk" is an entirely suitable name for what is problematic. Recent worries about possible mishaps from the use of recombinant DNA techniques in scientific research and industrial applications seem to me a case in which the term was accurately applied. But there is, in our time, a willingness to cluster an astonishingly large range of health, safety, and environmental problems under this one rubric. Thus, Douglas and Wildavsky, among other observers, rush to the conclusion that all environmentalists ever worry about are "risks." Misconceptions of that sort are ones we ought to avoid.

If the conservative voice in risk assessment were one that counseled us to cherish and preserve those parts of our culture, society, and environment that are genuinely worth saving, then I would, as a devoted conservative, view this work with much hope. Unfortunately, the style of conservatism we all too often find here is in the "damn the torpedoes" tradition of economic and technological thinking. Rather than lead us to devise better ways of expressing caution and care, such thinking often justifies a renewed recklessness.

Two kinds of issues in particular strike me as territory that ought to be rescued from this tendency. First is the issue of cases of actual harm—cancer, birth defects, other illnesses, deaths, damaged environments, and the like—obviously connected to profit-making industrial practices, yet sometimes treated as if their reality were merely probabilistic. We may visit the hospitals and gravesites, if we need to. We may wander through industrial wastelands and breathe deeply. But let us not pretend our troubles hinge on something like a gentlemanly roll of the dice or that other people's sickness and death can be deemed "acceptable" from some august, supposedly "neutral" standpoint. That only adds insult to injury.

Finally, there are times in which an issue that urgently needs to be addressed is poorly described (or not described at all) by subsuming it under the category of "risk." During the past decade or so a number of important items on the public agenda have been affected in this manner. Many of the interesting social problems linked to the development of nuclear power, for instance, have nothing to do with risk as such. But to a large extent the debate over risk and safety formed the range of public discussions on nuclear power to the exclusion of everything else.

A similar warping of the public's attention may well happen on another crucial question now before us. A technology recognized as an example of true risk in one context, rDNA research and development, will be badly misconstrued if it is seen as nothing more than a risk question in another emerging context—public policy on genetic engineering. It is one thing to think about the prospect that a lethal bug might escape from the laboratory and quite another to ponder what it means to assume direct control of the evolution of the human species. Possibilities made available by the new biotechnologies are profound ones. But they are not always questions of risk. It may happen, nevertheless, that because "risk" was the focus of discussions of rDNA research and development in the early years, it will continue to shape discussions on this topic for decades to come.[19] I could see, for example, the application of moral rules of the following kind: unless the development of a new genetic configuration can be shown to involve substantial quantifiable "risk," the development will be sanctioned for speedy implementation. If that happens, all the shortcomings of risk assessment will return to us with a vengeance.

* * *

"Didn't the fox *never* catch the rabbit, Uncle Remus," the little boy asked the next evening.

"He come mighty nigh it, honey, sho's you born."

Thus begins "The Wonderful Tar-Baby Story," an American classic told by Joel Chandler Harris. In the tale Brer Fox hatches a scheme to fool his arch rival, elusive Brer Rabbit. Using a mixture of tar and turpentine, he fashions a little figure, places it by the side of the road, and hides in the bushes to watch. Sure enough, before long Brer Rabbit comes pacing down the road and says good morning to the dark stranger. But "The Tar-Baby ain't sayin' nothin' and Brer Fox, he lay low."

Brer Rabbit takes the silence as an insult. After making repeated demands for a response, he hauls off and punches the Tar-Baby and gets his paw stuck. He tries another paw with the same result, then both hind feet, and finally his head.

"'Howdy, Brer Rabbit,' sez Brer Fox sezee. 'You look sorter stuck up dis mawning,' sezee, en den he rolled on de groun,' en laughed twel he couldn't laugh no mo'.'"

"Did the Fox eat the Rabbit?" asked the little boy who'd been listening to the story.

"Dat's all de fur de tale goes," replied the old man.[20]

9

BRANDY, CIGARS, AND HUMAN VALUES

Thou wall, O wall, O sweet and lovely wall,
Show me thy chink, to blink through with mine eyne!
    [Wall holds up his fingers.]
Thanks, courteous wall; Jove shield thee well for this!
But what see I? . . .
O wicked wall, through whom I see no bliss!
Curs'd be thy stones for deceiving me!

Snout in *Midsummer Night's Dream*
(5.01.176–183)

WE HAVE COME to the last resort. The search for reasonable moral limits to guide technological civilization needs a suitable climax. The concept of "nature" has turned out to be too ambiguous, too slippery to offer much guidance and the notion of "risk" too closely associated with unseemly gambling. The hour is late. Someone suggests I play my trump card. I must begin talking about "values."

In fact, the last resort is often a resort in the most literal sense: a place to vacation. The great minds of science, humanities, business, and university administration gather at a lovely retreat, a secluded country inn, a mansion beside a mountain lake, or a conference center by the sea. Fine food, elegant accommodations, cordial hosts, a crackling fire in the fireplace, and a well-organized program set the stage for what everyone knows will be a memorable occasion. If the guests grow tired of the panel discussions or paneled ceilings, they can always take a stroll through the forest or along the beach. What is the topic this time? "Science and Values"? "Values and the Future"? "Technology Assessment and Human Values"? "Values and the Professions"? Does it matter?

Often held up as a source of illumination on the most difficult questions and choices, the concept of "values" is better seen as a symptom of deep-seated confusion, an inability to think and talk precisely about the most basic questions of human well-being and the future of our planet. In a seemingly endless array of books, articles, and scholarly meetings, the hollow discourse about "values" usurps much of the space formerly occupied by much richer, more expressive categories of moral and political language. The longer such talk continues, the more vacuous it becomes, the further removed from any solid ground. "We begin to exchange set phrases," Nadezhda Mandelstam once observed, "not noticing that all living meaning has gone from them."[1]

Nowadays we are inclined to believe that there have always been "values" with a long history of discussion about them. But properly speaking, that is false. Of course, people have always had cares, commitments, responsibilities, preferences, tastes, religious convictions, personal aspirations, and so forth. But it is only in relatively recent times that anyone began to see all such things as "values" and lump them together within this one amorphous category.

The word "value," used as a noun, is, of course, a very old English term. It derives from the Latin *valere*, "to be strong, be of value." Throughout most of the history of its use, as recorded by the *Oxford English Dictionary*, the term meant "the worth of something," for example, the worth of an object in material exchange or the status or worthiness of a person in the eyes of others. In fact, the *Oxford English Dictionary* mentions no other definition for the noun other than variants of this simple meaning.

The term "value" enters social and political thought as a significant concept in the writings of eighteenth- and nineteenth-century political economists, most notably Adam Smith, David Ricardo, and later Karl Marx. In this context "value" meant the worth of a thing in an economic sense. Thus, a theory of value first appears as a development in economics. Later in the nineteenth century the idea of value was investigated by German historians and philosophers, Friedrich Nietzsche most prominently, who expanded and altered its meaning drastically. Here the term, usually mentioned in the plural, was understood to signify the sum total of principles, ideals, desires, and so forth that constitute the basic motivational structure of a person or, indeed, an entire culture. Nietzsche wrote ambitiously of the need for *Umwertung aller Werte*, the revaluation of all values. It was a task appropriate for modern scientific, industrial society which, in his view, had already succumbed to decadent nihilism.

An equally expansive but less pessimistic approach to the topic engaged pragmatist philosophers of the early twentieth century who sought a systematic approach to the study of "values," hoping it would become an important subspecialty within the study of ethics. Ralph Barton Perry proposed a "general theory of value," one that tried to give a reasonable account of the full range of human interests. Humanists and social scientists soon joined the discussion, many of them concerned with the supposed distinction between "facts" and "values." In a world of scientific fact, many of them asked, What are the proper boundaries for "value judgments"? To this day the question bothers social scientists who fear that surreptitious "values" may seep into and undermine otherwise pure empirical research. Others argue that "facts" cannot be separated from "values" in any clear, convenient way and that the attempt to maintain this boundary is a source of many errors in social inquiry. All empirical research rests upon value judgments about which facts are the important ones and how they ought to be studied. Values themselves, similarly, are factual matters rooted in empirical conditions of human living. Given this, the supposed distinction between facts and values falls apart.

Virtually unnoticed in the sparks and smoke arising from the controversy about "facts and values," however, is a fundamental shift in the meaning of the term "value" itself. Where once the term had a clear, simple, objective significance, it is now widely used in a psychological sense. Well into the nineteenth century "value" was taken to be *an attribute of something*. A person wanted, sought, used, or preserved a thing because it had worth or value. This sense of "value of" is, of course, still one of the proper uses of the term in our language. But it has now been joined by the linguistic fashion of using "value" to describe a purely *subjective phenomenon*, something in our heads as it were. We now say that individuals, groups, and societies "have values" that influence their activities. Values in this sense are something like general dispositions, which exist within us. Hence, we speak of "human values," "social values," "the values of the

middle class," "how modern science changes our values," and
so on.

The shift in the use of this term from an objective to a subjective meaning is strongly linked to a change in how we view our
situation. Raising the question of value is no longer so much an
occasion to think about the qualities of things or conditions outside us. Instead, it is an opportunity to look within, to perform
an inventory of emotions. Previously people saw themselves
pursuing certain kinds of activities because those activities had
value. Now we are more apt to conclude that *persons have values*
that lead them to behave in certain ways. This is a significant
change. It affects the very basis of our ideas about what is involved in human action.

Signs of this condition can be observed in the way those who
employ the concept interpret their own and others' subjectivity.
Speaking of "values" we are led to the almost inevitable conclusion that all such things are personal, arbitrary, irrational sentiments. You have your values, I have mine. Values are, we often
hear, mere preferences or "gut responses" and do not allow any
reasonable comparison. This view appeared as philosophical
doctrine in the writings of positivist philosophers earlier this
century. A. J. Ayer, for example, argued in *Language, Truth and
Logic* that all statements about value are deficient as compared to
statements about scientific fact because they are nothing more
than expressions of emotion. In subsequent philosophical debates that idea was roundly criticized by those who wished to
defend the place of reason and objectivity in ethics and politics.
But despite this effective counterattack in philosophy, the idea
still holds sway in many everyday discussions. People talk as if
the world were a kind of values supermarket in which everyone
has a shopping cart and selects what he or she wants in accor-
dance with internally held sentiments. Rational criticism at the
checkout counter is entirely inappropriate. How can one criticize subjective preferences?

Another important consequence of this way of talking and
thinking is to exclude much of what was formerly contained in
traditional moral and political language. That language includes
a rich multiplicity of categories, distinctions, arguments, modes
of justification, and areas of sensitivity that are simply overlooked in many contemporary conversations. The category of
"values" acts like a lawn mower that cuts flat whole fields of
meaning and leaves them characterless. Where previously we

might have talked about what was good, worthy, virtuous, or desirable, we are now reduced to speculating about "values." Where formerly it made sense to speak of rights and attempt to justify claims to those rights, one is inclined to moan plaintively about "values." Where not too long ago one could make a case for the wisdom of a particular action, one must now show how it matches someone's "values." Increasingly rare is the ability to make what were once fairly obvious distinctions and arguments. If there are any distinctions between an opinion and a principle, bias and belief, desire and need, one's individual interest and what one wishes for the community at large, people are less and less able to make them.

Perhaps the most important consequence of this state of affairs is a loss of attention paid to shared reasons for action. When values are looked upon as subjective, it makes little sense to raise questions of why or to ask for reasons. For the answer will inevitably be something like, "Because those are my values and that's why I'm doing what I'm doing." When basic moral and political ideas are bypassed with such alacrity, any hope for finding a rational basis for common action vanishes. Skill in making or understanding a complex moral argument becomes yet another of the lost arts. Max Horkheimer once described the decline of an objective sense of truth in the modern age as the "Eclipse of Reason." But what we encounter today is more accurately called an eclipse of reasons.

Examples of the condition I am describing afflict many debates about technology and public policy. Here discussions about "values" become part of attempts to achieve a systematic grasp of major issues of the day. The use of a new chemical or machine or technical process places various "options" before us. There are numerous "costs and benefits," "risks" and "impacts" that need to be taken into account. We can call upon existing scientific disciplines to provide theories and research findings relevant to an analysis of the issue. But what can help us weigh the various possibilities and come to an intelligent choice? The answer suddenly occurs to someone: social values must be understood in the context of "trade-offs." There will have to be compromise, give and take. Knowledge about people's values thus becomes an important "input" to the problem-solving, decision-making process.

But how is such knowledge to be obtained? The question has a bitter irony. For it involves nothing less than a desire to reinvent what has in fact already been discarded—a willingness to engage traditional ways of knowing and talking about the substance of social and political life. One solution often given is that social science can provide the missing link. Values can be surveyed, listed, measured, weighed, compared, and correlated to pave the way for our decisions. In the form of scientifically known "data" such values can be "plugged in" to help an otherwise directionless process find its way. Another solution frequently attempted is to employ certain "value experts" for their professional knowledge of such things. For this purpose, philosophers and other humanist intellectuals are often called upon to instruct engineers, planners, and decision makers as to what the "good" values are in our culture and how to maximize them.

Of course, approaches of this kind are doomed. The kind of knowledge policy deliberations require is not that which can be generated even by the most sophisticated of surveys, inventories, and weightings, provided by attitude research. The expectation that one can consult experts steeped in tradition, humanists able to deliver what others have forgotten, is a serious misunderstanding of what philosophers, historians, and students of literature have to offer. Those who accept such assignments and try to offer what Steven Marcus has called a "value fix" on our problems find themselves in a ludicrous position. Their role becomes much like that of Snout, the buffoon in Shakespeare's *Midsummer Night's Dream*, who holds up his fingers as the chink in the wall for others to look through.

Despite its vacuity as a concept (or perhaps because of it), the notion of "values" plays an important role in the bureaucratic and technocratic vocabulary. Its definition in that lexicon is something like "the residual concerns one needs to ponder after the real, practical business of society has been taken care of." Talking about or studying social issues needs to be done with the utmost caution; some of them may involve questions distressing to corporate or bureaucratic institutions. This is an especially serious problem when one looks to those institutions as one's source of funding. By invoking the hollow category of "values," however, one plays it safe. Government agencies fearful of the smallest signs of controversy can sponsor research on values without much fear of legislative criticism. Academic institutions with conservative faculties and administrators can hold conferences on politically troublesome topics if their sponsors shelter them under the mushy label "human values" rather than referring directly to such topics as social justice or the abuses of power.

An interdisciplinary organization of college teachers to which I belong, The Society for Religion in Higher Education, found a prudent way to solve a problem of this sort. During the middle 1970s its leaders sought ways to improve the society's image by giving it a more contemporary slant. They also hoped to make the society eligible for federal matching grants for its fund-raising drive and were advised that the term "religion" might cause trouble in view of the constitutional separation of church and state. The organization chose a convenient policy, changing its name to The Society for Values in Higher Education. After all, who could possibly object to "values" in higher education? What a masterstroke! And what a great tradition it upholds! The "Ten Values" Moses received from God on Mt. Sinai, Jesus of Nazareth's "Values on the Mount," St. Augustine's treatise *The City of Values*, Tom Paine's stirring pamphlet *The Values of Man*. And who will ever forget Karl Marx's magnum opus, *Capital and Social Values*.

An interesting feature of a conversation about values is that those who engage in it actually believe they are getting down to the most fundamental issues in human experience. In raising the topic they suppose they are moving away from purely rationalistic, technocratic modes of thinking toward something more warm, caring, and humane. An oft-repeated ritual manifests this care and indicates its level of importance. In the quiet of the evening after a delicious banquet, the scholars and administrators at a conference center by the sea gather for their peak experience. Out comes the brandy. Out come the cigars. And out comes the after-dinner speaker, an old trooper, usually a distinguished scientist or engineer, often someone who helped pioneer an advanced weapons system of some sort, to address his colleagues on "technology and human values." Typically, he observes that in our rush to achieve scientific and technical progress we have overlooked some very important questions. World hunger is still a dreadful problem. The environment is being damaged by human meddling. And nuclear weapons threaten to destroy all life on the planet. Somehow we need to restore our sense of proportion and make wise choices about how our science and technology are used. In talks such as these it is fashionable to make reference to "human" values, perhaps to distinguish them from the values representative of projects in which the speaker has been engaged until now. He was, it appears, too busy earlier in his career to take such issues seriously. At retirement, however, grave questions about "values" loom large. Having risen to the top of one's field, a person can begin to wonder about the fate of the planet and its suffering masses.

What we see in such instances are signs of thorough disorientation. Our institutions and professional careers are such that they simply do not encourage people to examine some of the most urgent human questions. That is why "values" appears as a hasty afterthought late in the day. In many instances the word "embarrassment" would serve just as well: How about "Science, Technology, and Our Embarrassment" as a title? It has a more honest ring. I would not hesitate to recommend it for those planning academic conferences or interdisciplinary programs for the undergraduate curriculum.

One obvious cure for the hollowness of "values" talk is to seek out terms that are more concrete, more specific. Whenever we feel the urge to say "human values" or "social values," perhaps we should immediately substitute a phrase closer to our intended meaning. If we mean "motives," then let's talk about them. If we mean "consumer preferences," then say so. If we mean "the norms of a particular group in society," then talk about those. If we mean "general moral principles that ought to guide our action," then explore, define, and defend those principles. What we will find, I believe, is that these more specific topics are an improvement over the vague label, and that once we've begun using them, the word "values" can never again substitute meaningfully for more substantial terms and questions.

But this superficial cure sidesteps the larger malady. A great many people, including some with considerable social power, seem to have lost the ability to link the specific, concrete conditions of their own work to any reasonable conception of human well-being. The question just never seems to come up. To remedy that would require a fundamental change in orientation for many organizations, vocations, and professions. We encourage people to become competent in a particular professional field, especially those concerned with inquiry into natural phenomena and the manipulation of material reality. At the same time we allow a scandalous incompetence in dealing with the fundamental, recurring questions of human existence: How are we to live together? How can we live gracefully and with justice? Questions of this nature are not, as some teachers like to tell their students, "soft" ones as compared to the "hard" research questions of science. They are as "hard" and as challenging as any that science could hope to tackle. They are, furthermore, eminently practical, involving the combined practice of ethics, politics, and technology.

By their very nature, however, such questions exclude expertise. I am one of those mistakenly called upon to perform in the ludicrous role of "value expert," and I can testify that the discussions and initiatives we need lie in no one's domain of specialized knowledge. The inquiry we need can only be a shared enterprise, a project of redemption that can and ought to include everyone.

Those who enter that conversation should draw upon the best resources available. Our words need the qualities of precision and depth, resources now threatened by the use of vacuous terms and haphazard thinking. A depleted language exacerbates many problems; a lively and concrete vocabulary offers the hope of renewal. We can still ask, How we are living now as compared to how we want to live? But let us not waste our time with "values." When you knock on that door, however loudly, no one answers.

10

THE WHALE AND
THE REACTOR

DIABLO CANYON in central California was not a place you could easily visit when I was growing up near there in the 1950s. Highway 101 turns inland at that point on the coastline and the road to Avila Beach did not extend northwest to that area. Unless one knew a rancher who owned land along that rugged shore, permission to travel to Diablo Canyon was hard to come by. Thus, one of the most beautiful and forbidding points at which the California mountains meet the sea was cut off from public view.

Technological progress, of course, has the ability to open up whole worlds previously shielded from our experience; the development of a scientific instrument or technical project makes it possible to see things we have never witnessed before. It was a visit to the construction site of one such technological wonder, a $5.5 billion nuclear power plant, that gave me the opportunity several years ago to view Diablo Canyon for the first time.

During a vacation in San Luis Obispo I decided to go out to the beaches where I spent my summers as a boy. As I drove along, I was attracted by a sign that announced Nuclear Tourist Information Center. I took the turn and soon found myself seated in a Pacific Gas and Electric Company tour bus listening to a young man talk about the power plant and environs in the upbeat voice used by tour guides from Williamsburg to Honolulu. "On a hill to your right you'll see the local Buddhist Church," he exclaimed. "I don't think it looks particularly oriental, do you?" He was referring to the temple that the Japanese citizens of the community had been able to build after fifteen years of hard work and saving following their release from the "relocation camps" of World War II. It was unsettling to hear the scenes of one's youth reinterpreted from a well-rehearsed, but largely ignorant public relations point of view.

As an example of excellence in highway engineering, the winding road to the reactor was fascinating. To handle the weight of the enormous equipment that had to pass over the hills, the roadbed had been laid almost completely level. It had none of the normal angular banking in its turns so that the trucks carrying heavy pieces of the plant would avoid tipping over. I noticed the guide taking special care with his driving, for the wrong momentum on any of the corners might send the bus plummeting into the sagebrush at the bottom of a ravine. As we made the half-hour journey to the site, we were regaled with stories about the economic benefits that the reactor would bring to the community, especially the many new jobs. But just after the guide had explained that virtue of the project he directed our attention to a construction facility at the side of the road, a totally automated cement-making factory which, he noted proudly, was operated by "only one man."

Feelings of anticipation swelled in me as the bus rolled to the top of the last hill separating us from our view of the Diablo Canyon site. As it reached the summit of a small plateau, I looked out over a vista that sent me reeling. Below us, nestled on the shores of a tiny cove, was the gigantic nuclear reactor, still under construction, a huge brown rectangular block and two white domes. In tandem the domes looked slightly obscene, like breasts protruding from some oversized goddess who had been carefully buried in the sand by the scurrying bulldozers. A string of electric cables suspended from high-energy towers ran downhill, awaiting their eventual connection to the power plant. In the waters just off shore lay two large rocks surrounded by a blanket of surf, as elegant in appearance as any that one can find along the Pacific coast. One of them, Diablo or "Devil's" Rock, loomed as a jagged pinnacle. Next to it, not too far away, was a smaller but even more finely sculptured piece of stone, Lion Rock, which looked very much like a lion at rest or, more accurately, like the Sphinx itself reclining on its haunches, paws outstretched on the surface of the ocean, silently asking its eternal question. At precisely that moment another sight caught my eye. On a line with the reactor and Diablo Rock but much farther out to sea, a California grey whale suddenly swam to the surface, shot a tall stream of vapor from its blow hole into the

165

air, and then disappeared beneath the waves. An overpowering silence descended over me.

In "The Virgin and the Dynamo" Henry Adams tells of an epiphany he experienced during his visit to the Great Exposition in Paris in 1900. As he gazed upon the forty-foot dynamo, Adams sensed something more than the power of the mechanical accomplishment: "the dynamo became a symbol of infinity." He began to feel the machine "as a moral force, much as the early Christians felt the Cross. . . . Before the end, one began to pray to it; inherited instinct taught the natural expression of man before the silent and infinite force." From these feelings of awe and mystery Adams was able to formulate a "law of acceleration" in human history, one that he believed could explain the increasingly complexity and rate of change in civilization.[1]

My own epiphany when confronted with the nuclear dynamo did not suggest anything as grand as a law of historical development. But that moment at Diablo Canyon crystallized a lifetime of experience. I suddenly realized how I had ended up at certain destinations that had long puzzled me.

A decade and a half earlier I had left San Luis Obispo to attend college. In the midst of my reading of history, politics, and philosophy I found myself drawn, quite unexpectedly, to questions concerning technology. I began to explore how changes in various technological systems affect the quality of personal and public life; how people make choices in technology as well as those instances in which they report that they have no choice, that things are "out of control"; how modern societies struggle with the question of limits; how social and political theorists have tried to come to terms with the dilemmas science and technology pose. This book is a product of that ongoing fascination.

Of course, the arguments, analyses, and observations anyone makes in his research and writing must stand or fall on their own merits. A writer must recognize normal standards of scholarly objectivity, give adequate evidence to back up assertions of fact, and proffer arguments in an unbiased, logical way. Observing such standards one says, in effect, what I offer is reliable; any reasonable person equipped with the same points of evidence and logic ought to arrive at similar conclusions.

But as important as these standards are, they leave two fundamental questions unanswered. Where does my own personal interest in these topics come from? Why have I chosen to approach them in the way I have? These are not questions scholars usually ask. A legitimate fear that one's work will be dismissed if it is afflicted by "subjectivity" leads many people to write with as little personal character or self-reference as possible. The "I" vanishes from the written page altogether. Personal experiences and judgments, even those relevant to the topic at hand are scrupulously avoided. The most intimate variety of self-reference that the prevailing style of scholarly writing allows is contained in the comforting phrase, "Our findings show that . . ." Yes, here are people we can trust.

But it seems to me that writers owe their readers something more. To be self-reflective, self-critical about the substance and method of one's thinking can help resolve puzzles that readers have about a work and its relationship to their own sense of things. I do not favor personal confessionals for their own sake, but, when all is said and done, books have intensely personal origins. Communities are shaped by the shared experiences of the people in them. To talk about these experiences directly may help reveal the ideas and sensibilities we hold in common.

There was a time when I thought the origins of my own interest in technology and politics could be found in my undergraduate years. Under the leadership of President Clark Kerr, The University of California had defined itself as a vast research, training, and social service "multiversity." Kerr had identified the new role of higher education as that of catering to the various organized interests in society who need scientific knowledge and trained personnel if they are to function efficiently. His views on the nature of modern society were outlined in a book, *Industrialism and Industrial Man*, which described the progress of industrialism as an overwhelming historical juggernaut that refashions the whole world to suit its purposes.[2] But his message took on a new significance in the fall of 1964 when, in the heat of the civil rights movement and the Johnson/Goldwater presidential campaign, the university banned all political advocacy on the Berkeley campus. In that light, Clark Kerr's eagerness to link the practices of education to the conditions of industrial society, that is, advanced capitalism, seemed completely reprehensible to those students who, like myself, had gone to college to discover the joy of learning, the life of the mind. For those of us involved in the demonstrations and heated debates of the Free Speech Movement, the issues were not only those of civil liberties and civil rights. A great many were also concerned about the character of institutions, such as ours, that seemed to have committed themselves to an authoritarian, technocratic path.

My personal misgivings at that time were soon amplified by experiences in the mid-1960s during two summers as a student intern at the Pentagon in Washington, D.C. On my first day at work, my boss, an army colonel, informed me that I was to become, of all things, a "systems analyst." From there on I received a steady diet of the techniques and accompanying jargon of such things as "planning, programming, and budgeting" and "program evaluation and review technique" that, under the leadership of Robert S. McNamara, were supposed to modernize and improve our nation's defense. The impressive rationality of it all stood in marked contrast to the senseless, bloody war then escalating rapidly in Southeast Asia. It became clear to me that an important function of the fancy computer models and convoluted technical language was sheer obfuscation. Decisions made on grounds of political expediency were justified as if judgments commanded by the most rigorous scientific analysis. I resolved to go on to graduate school to study the nature and significance of this strange new tendency in modern thought.

But as important as the embattled multiversity and Pentagon were in my undergraduate years, they were not, I later came to realize, the origin of my interest in technology and the modern predicament. The chance juxtaposition of the whale and reactor that day at Diablo Canyon opened my eyes to the fact that this fascination had much earlier sources. Here were two tangible symbols of the power of nature and of human artifice: one an enormous creature swimming gracefully in a timeless ecosystem, the other a gigantic piece of apparatus linked by sheer determination to the complicated mechanisms of the technological society. The first offered an image of things as they had always been, the other an image of things as they were rapidly coming to be. I realized that somehow I'd gotten caught in the middle. After long wanderings I'd returned to find the essence of the predicament I had been studying where I'd least expected it—in my own backyard.

My hometown, it is important to note, is situated almost exactly halfway between two large urban centers, San Francisco and Los Angeles, about two hundred and fifty miles from each. During the decade and a half that followed World War II the bucolic environment of San Luis Obispo was shaken again and again by technological and social transformations that seemed to emanate from one city or the other, or both. In a few short years the town witnessed the coming of freeways, supermarkets, jet airplanes, television, guided missiles (which I could watch from my front yard as they were shot from Vandenberg Air Force Base), computers, prefabricated houses in large tracts, wonder drugs, food additives, plastics, and a great many other innovations. The shape of the home and the activities of the family were refashioned to accommodate the arrival of all kinds of electronic gadgets. My parents still tell the story of buying a television set in 1953 in order to lure their two children back home, since my brother and I had gone off to watch Buck Rogers and the Cisco Kid on television at the neighbors' house across the street. I have vivid memories of the day I was playing in the huge field back of our house and was surprised to see a bulldozer begin pushing its way up the hill, the first piece of earth-moving equipment that eventually carved a four-lane freeway through the center of town. It was, in the most literal sense, the machine in the garden.

Many of the developments of those years were unquestionably beneficial. The Salk vaccine, for example, ended the horrible fear of what was then called "polio season," during which people had to be careful about going to the movies and other public places for fear of catching poliomyelitis. But many other innovations—changes in the texture of the natural and social landscape—were of dubious value and adopted with astonishing recklessness. It was this tendency of our California community to accept radical transformations of its way of life, to accept sweeping changes generated from far-away sources with little discussion, deliberation, or thought to the consequences that I now see as something truly astonishing. We were technological somnambulists wandering through an extended dream.

Much of this sleepwalking was, in retrospect, a product of pure innocence and ignorance. For instance, when I was a boy, my parents and I used to engage in a wonderful ritual of buying shoes. One of the things I liked best about it was that as you were being fitted, you walked over to a little machine, stepped onto a platform, and stared into an eyepiece that enabled you to take a long look right through the Buster Brown shoes and into your feet. It was a fluoroscope, a commercial X-ray machine, and we all thought it was just marvelous. Of course, you do not see the shoe store flouroscope anymore, and not because it's gone out of fashion. Children like me were being exposed to potentially hazardous amounts of radiation that might well have caused damage to the growth of bones in our feet. The machines operated at too high an intensity, were uncontrolled in the length of X-ray exposure, and actually provided little information that would assist in the fitting of shoes. I remember that I would sometimes gaze at the bones in my feet for minutes on end.

Of course, everything and everyone around me in those days tended to inspire a fast belief in the religion of progress. Modern was always thought to be superior to "old fashioned." The pattern, however, was certainly not one of tailoring technology to suit human needs. Instead, the practice was that of renovating human needs to match what modern science and engineering happened to make available. If there were problems encountered in following this path, they could be taken up later. But during this time the slogan was, in effect, "We don't know where we're going, but we're on our way."

In this light, another item that became exciting during the 1950s was ordinary soap. "Aren't you glad you use Dial Soap? Don't you wish everybody did?" the advertisement asked cheerfully. "It's the soap with hexachlorophene." Hexachlorophene is an antibacterial agent also added at the time to a cleanser called pHisoHex, which was thought to be especially good for the heartache of acne pimples. Hexachlorophene was used in these products as a matter of course for many years, but not any longer. In France in the early 1970s many infant deaths were linked to powders with high levels of hexachlorophene in them. At about the same time laboratory studies in the United States revealed that monkeys bathed in pHisoHex for five minutes daily over a period of time developed brain abnormalities. In September of 1972 the Food and Drug Administration required that hexachlorophene be available in prescription products only.

During the post–World War II era, in a sense all of us became unwitting subjects for a vast series of biological and social experiments, the results of which became apparent very slowly. Food additives—preservatives, coloring agents, and the like were added to the food we eat. Atmospheric nuclear bomb testing raised the level of radiation we encounter in a lifetime. Pesticides, herbicides, chemical fertilizers, and industrial pollutants were spread into the environment with abandon and now are present in our food and water supply. It should have been no surprise that the bad news about such substances as DDT,

PCB's, PBB's, and saccharin has dismayed people. "Everything causes cancer, so why bother." As a proclamation of defiance, many are prepared to say they don't care what the best scientific studies show about the possible carcinogenic properties of some artificial sweeteners: they want those sweeteners in their Coke and coffee anyway. They have gotten used to having the benefits of technological conveniences without expecting to pay the costs.

Of course, if anyone had bothered to notice, it should have been obvious that a price for "progress" was being paid all along. It was often a very subtle price, a barely recognizable price, but a real one nevertheless. In an earlier chapter I described it as a distinctive process, an alteration in forms of life. An entity well established in a particular sphere of practice, thoroughly useful, rich in its complex relationships within a given environment but not rigorously planned or rationally arranged, encounters a highly productive technology with a rationalized complexity of its own. More often than not the entity less well developed in an instrumental sense—the one less efficient, less productive—must yield. Its previous structure may make sense within the broader spectrum of life's activities but fail utterly under criteria used to judge the best available performance in a narrowly defined area of production. The structure of the entity—whether it be an implement, a commonly used product, a way of speaking, a way of thinking, a mode of activity, a pattern of work, a long-standing social institution, or custom—is interrupted and restructured to suit the requirements of the new arrangement of which it is to become a part. Or it is simply replaced by different apparatus and a different form of life. What is cut out in this process of increasing technical refinement, what is lopped off, inadvertently damaged, or permanently excluded may never reach anyone's awareness. Even if decision makers become aware of what is happening, their best calculations of cost/risk/benefit may indicate the price is well worth it. An urban renewal high rise is better than an old, decaying section of the city; prepackaged frozen foods are preferable to those prepared by traditional methods; computerized education is an advance over the old-fashioned methods of teaching and learning; the seven or eight hours Americans spend watching television each night are an improvement over whatever they did previously.

Such qualitative transformations of the human world present

a range of issues that are among the most interesting and most urgent of our times. Will our impressive scientific and technical powers produce a world genuinely superior to that which came before? Or will we be stuck with an accumulation of hare-brained, slipshod renovations that destroy far more than they improve? Questions of this kind are central to the most penetrating critiques of our technopolitan culture. But such issues are usually the very last ones to enter the minds of the businessmen, technical professionals, politicians, and others directly responsible for guiding the process of change. Increasingly, they place their faith in a range of techniques that promises practical results and quick profits while disregarding everything else. They know how to ask, Where's the bottom line? But they increasingly disregard a question each generation neglects at its peril, Where's the top line? What is the best of which our society is capable?

Reflecting on it now, I can identify many early experiences that eventually led me to conclude that when it comes right down to it, no one's minding the store. I remember that at a fairly young age I began wondering why it was that so many of the things presented as improvements were actually inferior to the things they replaced. In a wide variety of developments I watched from about 1950 onward, the claims made about modernization could not conceal the fact that pleasant, useful, and often beautiful *qualities* were simply missing from the new products. Attractive wooden buildings that showed the character of the human imagination and the carpenter's hand were replaced by flimsy stucco tract homes and apartment buildings. Land that was among the most fertile in the United States was paved over for highways, shopping centers, and motels. Even the roads, which in their primitive two-lane versions had paid some respect to the contours of the hills and placement of older houses and farms, with the coming of high-speed four-lane freeways began to slice directly through whatever stood in their paths: groves of ancient oak trees, whole neighborhoods, landmarks a hundred years old. "That's progress," people would say; or "You can't stop progress." They certainly did believe these changes were for the better, but they also believed that somehow the process was beyond their control. They were mere spectators, consumers of change.

One innovation typical of the ways in which new technical/rational processes affected the qualities of things could be noticed in the food. In a dairy community such as San Luis Obispo milk was delivered to each household's front door in glass bottles with cream floating on the top. Dairies in the area were especially proud of the creaminess of the milk they produced. Occasionally the newspaper would carry a story about a local cow that had broken the world's butter fat record. All of this, of course, was before the coming of the numerous refinements that have permanently changed the appearance, texture, and taste of milk products. In the practice dairy companies now follow, milk from the farm is taken to large processing centers that break the substance down into its basic components. Then, as the specific product requires, the various elements are resynthesized to make whole milk, nonfat milk, 2% low-fat milk, cheese, cottage cheese, ice cream, yogurt, etc. This is, I am sure, an extremely rational, efficient, and healthful way of doing things. But somehow the taste, texture, appearance, and complex combination of all those qualities shows something lacking in the products that undergo the process. Perhaps the most remarkable of all is the variety of ice cream which, through the miracles of modern chemistry, refuses to melt; it just sits there, gets warmer, and holds its shape.

Transformations such as those in the economic and technical organization of the dairy business were sometimes accompanied by advertising campaigns that were, in retrospect, as much a way of winning public support as they were to sell products. During an evening in the mid-1950s I sat on the floor in the middle of the living room watching television. All of a sudden there appeared on the screen a new commercial: vivid, slick, and impressive. It was a musical cartoon featuring, of all things, a black-and-white spotted cow with a broad smile on its face and wheels instead of hoofs. A catchy tune played in the background as it rolled over hill and glen on little railroad tracks whistling like a locomotive rather than mooing like a cow. The point of it all was to introduce the television audience to a powerful new force in the California economy: Foremost Dairies. Foremost, it turned out, had recently bought out a whole collection of local dairy concerns and was bringing us the new forms of mass production, mass distribution, cardboard throw-away containers, and that most peculiar of social fetishes, "convenience." I remember being completely thrilled by the whistling cow and proud to see that progressive company come to our town. And it was shortly thereafter that for reasons nobody seemed very clear about, milk products and dairying as a way of life changed for good.

In our society's enthusiasm to rationalize, standardize, and modernize, it has often thoughtlessly discarded qualities that it might, on more careful reflection, have wanted to preserve. Our institutions have engaged in a continuing process of reverse adaptation, in which things are reshaped to suit the technical means available. From almost every important sign it appears that this process still moves ahead relentlessly and without limit. Just as Henry Adams recognized a law of acceleration as he confronted the dynamo, my experience at Diablo Canyon suggested an obvious pattern of events. If there is a distinctive path that modern technological change has followed, it is that *technology goes where it has never been.* Technological development proceeds steadily from what it has already transformed and used up toward that which is still untouched. Thus, as one drives from Los Angeles to San Luis Obispo and on to San Francisco, one senses a certain inevitability in the way the landscape is going to change. People from Los Angeles and the Bay Area, from Boston and Newark, will go searching for greener pastures, for places to develop, places to retire. More and more of the countryside will be torn up, paved over, and "improved." Few people are capable of acknowledging an obvious fact: Americans now value countryside according to the extent to which it has not been despoiled by reckless economic and technological exploitation. Having ruined the natural and human ecology of the places they came from, they move on to begin the process anew. Of course, this process is not confined to geographical places. The gene pool itself is the obvious next target for innovation and development.

Everyone's hometown, I suppose, contains a monument or two to this remarkable progression. Mine just happens to be San Luis Obispo, one of the settlements founded by Father Junipero Serra during the late eighteenth century as he worked to secure coastal California for the Church and the Spanish Crown. In point of fact, Mission San Luis Obispo was the place of an interesting technical breakthrough that greatly enhanced Serra's accomplishments. Originally, the roofs of mission buildings were made of dry tule thatch. These, however, were easily torched by local Indians who fought the coming of the white man's religion, military troops, regimented labor, and other features of civilization. In 1790, after repeated Indian attacks and disastrous fires, the missionaries at San Luis Obispo discovered a way to eliminate the problem of flaming arrows, building roofs with locally produced tiles similar to ones they remembered from Spain. That was the beginning of the end of native American resistance in California. To be more accurate, it was a step toward the eventual extinction of those people. At present, the Vatican is studying Father Serra's qualifications to see if he will become the first North American priest to achieve sainthood.

*　*　*

Although I had known some of the details of the planning and construction of the Diablo Canyon reactor, I was truly shocked to see it actually sitting near the beach that sunny day in December. As the grey whale surfaced, it seemed for all the world to be asking, Where have you been? The answer was, of course, that I'd been in far-away places studying the moral and political dilemmas that modern technology involves, never imagining that one of the most pathetic examples was right in my hometown. My experience with the reactor itself, seeing it at a particular time and place, said infinitely more than all of the analyses and findings of all the detailed studies I had been reading ever could.

About that power plant, of course, the standard criticisms of nuclear power certainly hold. It does pose the dangers of catastrophic nuclear accident similar to the one at Three Mile Island. Certainly it will produce routine releases of low-level radiation and thermal pollution of the surrounding ocean water. No one has developed a coherent plan for storing the long-lived radioactive wastes that this plant and others like it will generate. Already ten times more costly than originally estimated, its $5.5 billion price tag does not include the tens of millions of dollars it will cost to decommission the thing when its working life has ended.

From the point of view of civil liberties and political freedom, Diablo Canyon is a prime example of an inherently political technology. Its workings require authoritarian management and extremely tight security. It is one of those structures, increasingly common in modern society, whose hazards and vulnerability require them to be well policed. What that means, of course, is that insofar as we have to live with nuclear power, we ourselves become increasingly well policed.

Sophisticated arguments pro and con on issues of this kind have involved some of the best minds in America. I take these analyses and quantitative demonstrations very seriously. How

175

can one do otherwise? After all, one of the reasons Diablo Canyon was chosen as a site for a nuclear power plant was that studies showed it to be one of the few places on the California coast free from the danger of earthquakes. Even some environmentalist organizations approved the site as an exemplary choice for that reason. Unfortunately, later geological findings proved the conclusion erroneous; there is actually an active earthquake fault located just offshore. Researchers have now turned to a fascinating new question, How large will the earthquakes be?[3]

But beyond the sophisticated studies of scientists and policy analysts concerned with these issues lies another consideration, which, if we ever become incapable of recognizing it, will indicate that our society has lost its bearings, that it is prepared to feed everything into the shredder. To put the matter bluntly, in that place, on that beach, against those rocks, mountains, sands, and seas, the power plant at Diablo Canyon is simply a hideous mistake. It is out of place, out of proportion, out of reason. It stands as a permanent insult to its natural and cultural surroundings. The thing should never have been put there, regardless of what the most elegant cost/benefit, risk/benefit calculations may have shown. Its presence is a tribute to those who cherish power and profit over everything in nature and our common humanity.

To write any such conclusion in our time is, I realize, virtual heresy. My colleagues in the science, technology, and society assessment business often counsel me to be more careful, to reword my point of view to say, "We must now consider the possibility that in proceeding along this path we risk losing something of inestimable value that we may later begin to miss." But the plant and its inherent destructiveness are already much more than "possibilities." They are already in place. More and more the whole language used to talk about technology and social policy—the language of "risks," "impacts," and "trade-offs"—smacks of betrayal. The excruciating subtleties of measurement and modeling mask embarrassing shortcomings in human judgment. We have become careful with numbers, callous with everything else. Our methodological rigor is becoming spiritual rigor mortis.

At this writing construction of the first of two nuclear reactors at Diablo Canyon is finished. The Nuclear Regulatory Commission has granted a full power license and the plant has begun generating electricity for the first time. Discovery of the nearby earthquake fault along with embarrassing disclosures of structural flaws and shoddy workmanship forced a thorough, enormously costly rebuilding of the plant during the last several years of the project. Some of the plant's engineers and workers still insist that the structure is unsound and complain that they have been harassed on the job for talking to the press about these troubles. To recoup the spiraling costs of the installation, the Pacific Gas and Electric Company has already applied to the Public Utilities Commission for the first of what is bound to be a series of steep rate increases.

During public protests of the early 1980s, several thousand demonstrators were arrested at the gates of Diablo Canyon. With the completion of the plant and its licensing by the Nuclear Regulatory Commission, the protests have dwindled. A number of citizens' action groups have vowed to exhaust every legal alternative in opposing the reactor's operation through the courts. Always sensitive to public fears and pressures, Pacific Gas and Electric has renamed its visitors' bureau near Avila Beach the Energy Information Center, cleverly deleting all reference to the word "nuclear." George Orwell's Newspeak could have done no better. While the conservative-minded residents of San Luis Obispo and surrounding communities are reluctant to protest or file suit, a great many of them feel they have been swindled and endangered by the coming of this ghastly new neighbor. In every sense short of a meltdown, Diablo Canyon is already a disaster.

Two years after my epiphany I was invited back to my hometown to give a lecture on technology and the environment. During the talk I argued that while Diablo Canyon was not a very good place for a reactor, it would still be a wonderful spot for a public park. Rather than start up the plant, I suggested, the community ought to take it over, dedicating it as a monument to the nuclear age. For generations thereafter parents could take their children to the place, have leisurely picnics on the beach, and think back to the time when we finally came to our senses. Of course, my modest proposal was not taken seriously; too much money had already been spent, too much institutional momentum built up, too many careers invested, too many sermons preached from the pulpits of progress to allow any course of action that sensible. At present our society seems to prefer monuments of a different kind—monuments to gigantism, war,

177

and the overstepping of natural and cultural boundaries. Such are the accomplishments we support with our dollars and our votes. How long will it be until we are ready for anything better?

NOTES

CHAPTER 1. *Technologies as Forms of Life*

1. Tom Wolfe, *The Right Stuff* (New York: Bantam Books, 1980), 270.
2. *The Encyclopedia of Philosophy*, 8 vols., Paul Edwards (editor-in-chief) (New York: Macmillan: 1967).
3. *Bibliography of the Philosophy of Technology*, Carl Mitcham and Robert Mackey (eds.) (Chicago: University of Chicago Press, 1973).
4. There are, of course, exceptions to this general attitude. See Stephen H. Unger, *Controlling Technology: Ethics and the Responsible Engineer* (New York: Holt, Rinehart and Winston, 1982).
5. An excellent corrective to the general thoughtlessness about "making" and "use" is to be found in Carl Mitcham, "Types of Technology," in *Research in Philosophy and Technology*, Paul Durbin (ed.) (Greenwich, Conn. JAI Press, 1978), 229–294.
6. J. L. Austin, *Philosophical Papers* (Oxford: Oxford University Press, 1961), 123–152.
7. See William Kolender et al., "Petitioner v. Edward Lawson," *Supreme Court Reporter* 103: 1855–1867, 1983. Edward Lawson had been arrested approximately fifteen times on his long walks and refused to provide identification when stopped by the police. Lawson cited his rights guaranteed by the Fourth and Fifth Amendments of the U.S. Constitution. The Court found the California vagrancy statute requiring "credible and reliable" identification to be unconstitutionally vague. See also Jim Mann, "State Vagrancy Law Voided as Overly Vague," *Los Angeles Times*, May 3, 1983, 1, 19.
8. Ludwig Wittgenstein, *Philosophical Investigations*, ed. 3, translated by G. E. M. Anscombe, with English and German indexes (New York: Macmillan, 1958), 11e.
9. Hanna Pitkin, *Wittgenstein and Justice: On the Significance of Ludwig Wittgenstein for Social and Political Thought* (Berkeley: University of California Press, 1972), 293.
10. For a thorough discussion of this idea, see Hannah Arendt, *The Human*

Condition (Chicago: University of Chicago Press, 1958); and Hannah Arendt, *Willing,* vol II of *The Life of the Mind* (New York: Harcourt Brace Jovanovich, 1978).

11. *Philosophical Investigations,* 226e.

12. Karl Marx and Friedrich Engels, "The German Ideology," in *Collected Works,* vol. 5 (New York: International Publishers, 1976), 31.

13. Karl Marx, *Grundrisse,* translated with a foreword by Martin Nicolaus (Harmondsworth, England: Penguin Books, 1973), 325.

14. An interesting discussion of Marx in this respect is Kostas Axelos' *Alienation, Praxis and Technē in the Thought of Karl Marx,* translated by Ronald Bruzina (Austin: University of Texas Press, 1976).

15. Karl Marx, *Capital,* vol. 1, translated by Ben Fowkes, with an introduction by Ernest Mandel (Harmondsworth, England: Penguin Books, 1976), chap. 15.

16. *Philosophical Investigations,* 49e.

CHAPTER 2. *Do Artifacts Have Politics?*

1. Lewis Mumford, "Authoritarian and Democratic Technics," *Technology and Culture* 5:1–8, 1964.

2. Denis Hayes, *Rays of Hope: The Transition to a Post-Petroleum World* (New York: W. W. Norton, 1977), 71, 159.

3. David Lillienthal, *T.V.A.: Democracy on the March* (New York: Harper and Brothers, 1944), 72–83.

4. Daniel J. Boorstin, *The Republic of Technology* (New York: Harper and Row, 1978), 7.

5. Langdon Winner, *Autonomous Technology: Technics-Out-of-Control as a Theme in Political Thought* (Cambridge: MIT Press, 1977).

6. The meaning of "technology" I employ in this essay does not encompass some of the broader definitions of that concept found in contemporary literature, for example, the notion of "technique" in the writings of Jacques Ellul. My purposes here are more limited. For a discussion of the difficulties that arise in attempts to define "technology," see *Autonomous Technology,* 8–12.

7. Robert A. Caro, *The Power Broker: Robert Moses and the Fall of New York* (New York: Random House, 1974), 318, 481, 514, 546, 951–958, 952.

8. Robert Ozanne, *A Century of Labor-Management Relations at McCormick and International Harvester* (Madison: University of Wisconsin Press, 1967), 20.

9. The early history of the tomato harvester is told in Wayne D. Rasmussen, "Advances in American Agriculture: The Mechanical Tomato Harvester as a Case Study," *Technology and Culture* 9:531–543, 1968.

10. Andrew Schmitz and David Seckler, "Mechanized Agriculture and Social Welfare: The Case of the Tomato Harvester," *American Journal of Agricultural Economics* 52:569–577, 1970.

11. William H. Friedland and Amy Barton, "Tomato Technology," *Society* 13:6, September/October 1976. See also William H. Friedland, *Social Sleep-*

walkers: Scientific and Technological Research in California Agriculture, University of California, Davis, Department of Applied Behavioral Sciences, Research Monograph No. 13, 1974.

12. *University of California Clip Sheet* 54:36, May 1, 1979.

13. "Tomato Technology."

14. A history and critical analysis of agricultural research in the land-grant colleges is given in James Hightower, *Hard Tomatoes, Hard Times* (Cambridge: Schenkman, 1978).

15. David F. Noble, *Forces of Production: A Social History of Machine Tool Automation* (New York: Alfred A. Knopf, 1984).

16. Friedrich Engels, "On Authority," in *The Marx-Engels Reader,* ed. 2, Robert Tucker (ed.) (New York: W. W. Norton, 1978), 731.

17. Ibid.

18. Ibid., 732, 731.

19. Karl Marx, *Capital,* vol. 1, ed. 3, translated by Samuel Moore and Edward Aveling (New York: Modern Library, 1906), 530.

20. Jerry Mander, *Four Arguments for the Elimination of Television* (New York: William Morrow, 1978), 44.

21. See, for example, Robert Argue, Barbara Emanuel, and Stephen Graham, *The Sun Builders: A People's Guide to Solar, Wind and Wood Energy in Canada* (Toronto: Renewable Energy in Canada, 1978). "We think decentralization is an implicit component of renewable energy; this implies the decentralization of energy systems, communities and of power. Renewable energy doesn't require mammoth generation sources of disruptive transmission corridors. Our cities and towns, which have been dependent on centralized energy supplies, may be able to achieve some degree of autonomy, thereby controlling and administering their own energy needs." (16)

22. Alfred D. Chandler, Jr., *The Visible Hand: The Managerial Revolution in American Business* (Cambridge: Belknap, 1977), 244.

23. Ibid.

24. Ibid., 500.

25. Leonard Silk and David Vogel, *Ethics and Profits: The Crisis of Confidence in American Business* (New York: Simon and Schuster, 1976), 191.

26. Russell W. Ayres, "Policing Plutonium: The Civil Liberties Fallout," *Harvard Civil Rights—Civil Liberties Law Review* 10 (1975): 443, 413–414, 374.

CHAPTER 3. *Technē and Politeia*

1. Plato, *Laws,* 7.803b, translated by A. E. Taylor, in *The Collected Dialogues of Plato,* Edith Hamilton and Huntington Cairns (eds.) (Princeton, N.J.: Princeton University Press, 1961), 1374.

2. Jean-Jacques Rousseau, *The Social Contract,* translated and introduced by Maurice Cranston (New York: Penguin Books, 1968), 84.

3. Alexander Hamilton, "Federalist No. 9," in *The Federalist Papers,* with an introduction by Clinton Rossiter (New York: Mentor Books, 1961), 72–73.

4. Thomas Jefferson, "Notes on Virginia," in *The Life and Selected Writings of Thomas Jefferson*, Adrienne Koch and William Peden (eds.) (New York: Modern Library, 1944), 28–281.

5. Quoted in Gordon S. Wood, *The Creation of the American Republic, 1776–1787* (New York: W. W. Norton, 1972), 114.

6. See John Kasson, *Civilizing the Machine: Technology and Republican Values in America, 1776–1900* (New York: Grossman, 1976).

7. See Albert O. Hirschman, *The Passions and the Interests: Political Arguments for Capitalism Before Its Triumph* (Princeton: Princeton University Press, 1977).

8. Denison Olmsted, "On the Democratic Tendencies of Science," *Barnard's Journal of Education*, vol. 1, 1855–1856, reprinted in *Changing Attitudes Toward American Technology*, Thomas Parke Hughes (ed.) (New York: Harper and Row, 1975), 148.

9. Karl Polanyi, *The Great Transformation* (Boston: Beacon Press, 1957), 33.

10. "Beta" (Edward W. Byrn), "The Progress of Invention During the Past Fifty Years," *Scientific American*, vol. 75, July 25, 1896, reprinted in Hughes (see note 8), 158–159.

11. For a discussion of the wide-ranging influence of the idea of efficiency in American thought, see Samuel Haber, *Efficiency and Uplift: Scientific Management in the Progressive Era, 1890–1920* (Chicago: Chicago University Press, 1964); and Samuel P. Hays, *Conservation and the Gospel of Efficiency: The Progressive Conservation Movement, 1890–1920* (New York: Atheneum, 1959).

12. See William L. Kahrl, *Water and Power: The Conflict Over Los Angeles' Water Supply in the Owens Valley* (Berkeley: University of California Press, 1982).

13. An analysis of the efficacy of various kinds of government regulation is contained in *The Politics of Regulation*, James Q. Wilson (ed.) (New York: Basic Books, 1980).

14. A representative selection of the ambitious energy studies done during the 1970s would include the following: *A Time to Choose*, The Energy Policy Project of the Ford Foundation (Cambridge: Ballinger, 1974); *Nuclear Power: Issues and Choices*, Report of the Nuclear Energy Policy Study Group (Cambridge: Ballinger, 1977); *Energy in America's Future: The Choices Before Us*, National Energy Studies Project (Baltimore: Johns Hopkins University Press, 1979); *Energy: The Next Twenty Years*, Report of the Study Group Sponsored by the Ford Foundation and Administered by Resources for the Future, Hans Landsberg et al. (Cambridge: Ballinger, 1979); *Energy in Transition, 1985–2010*, National Research Council Committee on Nuclear and Alternative Energy Systems (San Francisco: W. H. Freeman, 1980).

15. *Nuclear Power: Issues and Choices*, 44, 41.

16. Amory B. Lovins, "Technology Is the Answer! (But What Was the Question?): Energy as a Case Study of Inappropriate Technology," discussion paper for the Symposium on Social Values and Technological Change in an

International Context, Racine, Wis., June 1978, 1.

17. Amory B. Lovins, *Soft Energy Paths: Toward a Durable Peace* (Cambridge: Ballinger, 1977), 6–7.

18. Paul Goodman, *People or Personnel and Like a Conquered Province* (New York: Vintage Press, 1968), 4.

19. See, for example, J. L. Smith, "Photovoltaics," *Science* 212:1472–1478, 1981.

CHAPTER 4. *Building the Better Mousetrap*

1. The best general introduction to appropriate technology in this context is Nicholas Jéquier, *Appropriate Technology: Problems and Promises* (Paris: Development Center of the OECD, 1976), especially Jéquier's essay, "The Major Policy Issues." A survey of activities in developing nations is given in Ken Darrow and Rick Pam, *Appropriate Technology Sourcebook*, 2 vols. (Stanford, Calif.: Volunteers in Asia, 1976 and 1981).

2. E. F. Schumacher, *Small is Beautiful: Economics as if People Mattered* (New York: Harper and Row, 1973), 31.

3. A survey of appropriate technology research and development in the United States can be found in *Appropriate Technology—A Directory of Activities and Projects*, Integrative Design Associates (Washington, D.C.: U.S. Government Printing Office, 1977).

4. See, for example, Robert Owen, *A New View of Society and a Report to the Country of Lanark*, with an introduction by V. A. C. Gatrell (ed.) (New York: Penguin Books, 1970); Peter Kropotkin, *Fields, Factories and Workshops* (Boston: Houghton Mifflin, 1899); Ralph Borsodi, *Flight from the City* (New York: Harper and Row, 1933); Peter van Dresser, *A Landscape for Humans* (Albuquerque: Biotechnic Press, 1972); and Murray Bookchin, *The Spanish Anarchists: The Heroic Years 1868–1936* (New York: Free Life Editions, 1977).

5. An extremely insightful treatment of this period is contained in Jo Durden-Smith, *Who Killed George Jackson?* (New York: Alfred A. Knopf, 1976).

6. Perhaps the best expression of his tendency is to be found in Karl Hess, *Dear America* (New York: Morrow, 1975), chaps. 8–10 and Hess's *Community Technology* (New York: Harper & Row, 1979).

7. *The Movement Toward a New America: The Beginnings of a Long Revolution*, assembled by Mitchell Goodman (New York: Alfred A. Knopf, 1970); and *The Whole Earth Catalog*, Steward Brand (ed.) and colleagues (Menlo Park, Calif.: Nowels, 1968).

8. See *Weatherman*, Harold Jacobs (ed.) (New York: Ramparts Press, 1970), for expressions of this view.

9. *The Whole Earth Catalog*, 1.

10. Lewis Mumford, *The Myth of the Machine: The Pentagon of Power* (New York: Harcourt Brace Jovanovich, 1970); Paul Goodman, *The New Reformation: Notes of a Neolithic Conversative* (New York: Random House, 1970); Herbert

Marcuse, *One-Dimensional Man: Studies in the Ideology of Advanced Industrial Society* (Boston: Beacon Press, 1964); Theodore Roszak, *The Making of a Counter Culture; Reflections on Technocratic Society and Its Youthful Opposition* (Garden City, N.Y.: Doubleday, 1969); and Jacques Ellul, *The Technological Society*, translated by John Wilkinson (New York: Alfred A. Knopf, 1964).

11. For an exposition of the central notions in this tradition, see Langdon Winner, *Autonomous Technology: Technics-out-of-Control as a Theme in Political Thought* (Cambridge: MIT Press, 1977).

12. Thomas Carlyle, "Signs of the Times," in *Selected Works, Reminiscences and Letters*, Julian Symons (ed.) (Cambridge: Harvard University Press, 1957), 25.

13. Herbert Marcuse, *An Essay on Liberation* (Boston: Beacon Press, 1969), 19.

14. Ibid., 90.

15. *The Making of a Counter Culture*, chap. 6.

16. Theodore Roszak, *Where The Wasteland Ends: Politics and Transcendence in Postindustrial Society* (Garden City, N.Y.: Anchor Books, 1973), 396.

17. Ibid., 394.

18. See any issue of *The Review of Radical Political Economics* for the directions some American economists have taken here.

19. *The Journal of the New Alchemists* 1 (1973):50.

20. Ibid., see inside front cover.

21. Amory B. Lovins, *Soft Energy Paths: Toward a Durable Peace* (Cambridge: Ballinger, 1977). See also Lovins' "A Neo-Capitalist Manifesto: Free Enterprise Can Finance Our Energy Future," *Politicks and Other Human Interests* 1(2):15–18, 1978.

22. David Dickson, *The Politics of Alternative Technology* (New York: Universe Books, 1975).

23. Murray Bookchin, *Post-Scarcity Anarchism* (Berkeley: Ramparts Press, 1971). Bookchin has since published his magnum opus, *The Ecology of Freedom: The Emergence and Dissolution of Hierarchy* (Palo Alto: Cheshire Books, 1982).

24. Richard Eckaus, *Appropriate Technology for Developing Countries* (Washington, D.C., National Academy of Sciences, 1977). Unsympathetic to the idea of appropriate technology from the outset, the book's title is something of a hoax.

25. Quoted in *The Politics of Alternative Technology*, 102, 103–104.

26. Victor Papanek and James Hennessey, *How Things Don't Work* (New York: Pantheon Books, 1977), 49.

27. Ibid., 119.

28. Ibid., xiii.

29. Ibid., 27.

30. The saying, attributed to an address Emerson gave, appears in *Borrowings: A Collection of Beautiful Thoughts Selected from Great Authors*, compiled by Sarah S. B. Yule and Mary S. Keene (New York: Dodge, 1889).

31. Arthur Eugene Bestor, Jr., *Backwoods Utopias: The Sectarian and Owenite Phases of Communitarian Socialism in America, 1663–1829* (Philadelphia: University of Pennsylvania Press, 1950). See also Dolores Hayden, *Seven American Utopias: The Architecture of Communitarian Socialism* (Cambridge: MIT Press, 1976).

32. This emphasis is readily apparent in *Rainbook: Resources for Appropriate Technology* (New York: Schocken Books, 1977).

33. Typical of the attacks on appropriate technology from the left is Alexander Cockburn and James Ridgeway's "The Myth of Appropriate Technology," *Politicks and Other Human Interests* 1(4): 1977. "Technology," they argue, "whether alternative or intermediate, or even just big, can only exist as a reflection of the existing political and economic system. Appropriate technology, per se, won't change anything. A politics that may include some aspects of smaller technology may lead to change—but the politics must come first." (28)

34. Harvey Wasserman, *American Born and Reborn* (New York: Collier Books, 1983), 303–304.

35. For a lively defense of the Greens, see Fritjof Capra and Charlene Spretnak, *Green Politics* (New York: E. P. Dutton, 1984).

36. Bertolt Brecht, "The Exception and the Rule," in *The Jewish Wife and Other Short Plays*, English versions by Eric Bentley (New York: Grove Press, 1965), 111.

CHAPTER 5. *Decentralization Clarified*

1. John Naisbitt, *Megatrends: Ten New Directions Transforming Our Lives* (New York: Warner Books, 1984), 103.

2. Robert A. Caro, *The Power Broker: Robert Moses and the Fall of New York* (New York: Alfred A. Knopf, 1974), 753.

3. See the *Oxford English Dictionary* for early uses of these terms in print.

4. Paul Goodman, *Drawing the Line*, Taylor Stoehr (ed.) (New York: New Life, 1978), 185. Goodman's thoughts on decentralism are argued in detail in *People or Personnel and Like a Conquered Province* (New York: Vintage Press, 1968).

5. Peter Kropotkin, *Mutual Aid, a Factor of Evolution* (London: William Heinemann, 1902); and *Fields, Factories and Workshops* (Boston: Houghton Mifflin, 1899).

6. Murray Bookchin, *Post-Scarcity Anarchism* (Palo Alto: Ramparts Press, 1970); *The Spanish Anarchists: The Heroic Years, 1868–1936* (New York: Free Life Editions, 1977); *Toward an Ecological Society* (Montreal: Black Rose Books, 1981); and *The Ecology of Freedom: The Emergence and Dissolution of Hierarchy* (Palo Alto: Cheshire Books, 1982).

7. G. D. H. Cole, *Guild Socialism Re-Stated* (London: Leonard Parsons, 1920), 13.

8. Ibid., 61.

9. Joseph K. Hart, "Power and Culture," reprinted in *Changing Attitudes Toward American Technology*, Thomas Parke Hughes (ed.) (New York: Harper and Row, 1975), 250, originally published in *The Survey*, Graphic Number 51, 625–628, March 1, 1924.

CHAPTER 6. *Mythinformation*

1. See, for example, Edward Berkeley, *The Computer Revolution* (New York: Doubleday, 1962); Edward Tomeski, *The Computer Revolution: The Executive and the New Information Technology* (New York: Macmillan, 1970); and Nigel Hawkes, *The Computer Revolution* (New York: E. P. Dutton, 1972). See also Aaron Sloman, *The Computer Revolution in Philosophy* (Hassocks, England: Harvester Press, 1978); Zenon Pylyshyn, *Perspectives on the Computer Revolution* (Englewood Cliffs, N.J.: Prentice-Hall, 1970); Paul Stoneman, *Technological Diffusion and the Computer Revolution* (Cambridge: Cambridge University Press, 1976); and Ernest Braun and Stuart MacDonald, *Revolution in Miniature: The History and Impact of Semiconductor Electronics* (Cambridge: Cambridge University Press, 1978).

2. Michael L. Dertouzos in an interview on "The Today Show," National Broadcasting Company, August 8, 1983.

3. Daniel Bell, "The Social Framework of the Information Society," in *The Computer Age: A Twenty Year View*, Michael L. Dertouzos and Joel Moses (eds.) (Cambridge: MIT Press, 1980), 163.

4. See, for example, Philip H. Abelson, "The Revolution in Computers and Electronics," *Science* 215:751–753, 1982.

5. Edward A. Feigenbaum and Pamela McCorduck, *The Fifth Generation: Artificial Intelligence and Japan's Computer Challenge to the World* (Reading, Mass.: Addison-Wesley, 1983), 8.

6. Among the important works of this kind are David Burnham, *The Rise of the Computer State* (New York: Random House, 1983); James N. Danziger et al., *Computers and Politics: High Technology in American Local Governments* (New York: Columbia University Press, 1982); Abbe Moshowitz, *The Conquest of Will: Information Processing in Human Affairs* (Reading, Mass.: Addison-Wesley, 1976); James Rule et al., *The Politics of Privacy* (New York: New American Library, 1980); and Joseph Weizenbaum, *Computer Power and Human Reason: From Judgment to Calculation* (San Francisco: W. H. Freeman, 1976).

7. Quoted in Jacques Vallee, *The Network Revolution: Confessions of a Computer Scientist* (Berkeley: And/Or Press, 1982), 10.

8. Tracy Kidder, *Soul of a New Machine* (New York: Avon Books, 1982), 228.

9. *The Fifth Generation*, 14.

10. James Martin, *Telematic Society: A Challenge for Tomorrow* (Englewood Cliffs, N.J.: Prentice-Hall, 1981), 172.

11. Ibid., 4.

12. Starr Roxanne Hiltz and Murray Turoff, *The Network Nation: Human*

Communication via Computer (Reading, Mass.: Addison-Wesley, 1978), 489.

13. Ibid., xxix.

14. Ibid., 484.

15. Ibid., xxix.

16. *The Network Revolution*, 198.

17. John Naisbitt, *Megatrends: Ten New Directions Transforming Our Lives* (New York: Warner Books, 1984), 282.

18. Amitai Etzioni, Kenneth Laudon, and Sara Lipson, "Participating Technology: The Minerva Communications Tree," *Journal of Communications*, 25:64, Spring 1975.

19. J. C. R. Licklider, "Computers and Government," in Dertouzos and Moses (eds.), *The Computer Age*, 114, 126.

20. Quoted in *The Fifth Generation*, 240.

21. *Occupational Outlook Handbook, 1982–1983*, U.S. Bureau of Labor Statistics, Bulletin No. 2200, Superintendent of Documents, U.S. Government Printing Office, Washington, D.C. See also Gene I. Maeroff, "The Real Job Boom Is Likely to be Low-Tech," *New York Times*, September 4, 1983, 16E.

22. See, for example, James Danziger et al., *Computers and Politics*.

23. For a study of the utopia of consumer products in American democracy, see Jeffrey L. Meikle, *Twentieth Century Limited: Industrial Design in America, 1925–1939* (Philadelphia: Temple University Press, 1979). For other utopian dreams see Joseph J. Corn, *The Winged Gospel: America's Romance with Aviation, 1900–1950* (Oxford: Oxford University Press, 1983); Joseph J. Corn and Brian Horrigan, *Yesterday's Tomorrows: Past Visions of America's Future* (New York: Summit Books, 1984); and Erik Barnow, *The Tube of Plenty* (Oxford: Oxford University Press, 1975).

24. *The Fifth Generation*, 8.

25. "The Philosophy of US," from the official program of The US Festival held in San Bernardino, California, September 4–7, 1982. The outdoor rock festival, sponsored by Steven Wozniak, co-inventor of the original Apple Computer, attracted an estimated half million people. Wozniak regaled the crowd with large-screen video presentations of his message, proclaiming a new age of community and democracy generated by the use of personal computers.

26. "*Power* corresponds to the human ability not just to act but to act in concert. Power is never the property of an individual; it belongs to a group and remains in existence only so long as the group keeps together." Hannah Arendt, *On Violence* (New York: Harcourt Brace & World, 1969), 44.

27. John Markoff, "A View of the Future: Micros Won't Matter," *Info-World*, October 31, 1983, 69.

28. Donald H. Dunn, "The Many Uses of the Personal Computer," *Business Week*, June 23, 1980, 125–126.

CHAPTER 7. *The State of Nature Revisited*

1. Ralph Waldo Emerson, "Nature," in *The Selected Writings of Ralph Waldo Emerson*, Brooks Atkinson (ed.) (New York: Modern Library, 1940).

2. General introductions to changing conceptions of nature in Western culture include Clarence J. Glacken, *Traces on the Rhodian Shore* (Berkeley: University of California Press, 1967); John Passmore, *Man's Responsibility for Nature* (London: Gerald Duckworth, 1974); and R. G. Collingwood, *The Idea of Nature* (Oxford: Clarendon Press, 1954). Discussions of the idea in American history include Perry Miller, *Errand Into the Wilderness* (Cambridge: Belknap, 1956); Hans Huth, *Nature and the American: Three Centuries of Changing Attitudes* (Berkeley: University of California Press, 1957); Roderick Nash, *Wilderness and the American Mind*, revised edition (New Haven: Yale University Press, 1973); and Carroll Pursell (ed.), *From Conservation to Ecology: The Development of Environmental Concern* (New York: T. Y. Crowell, 1973).

3. Thomas Hobbes, *Leviathan or the Matter, Forme and Power of a Commonwealth Ecclesiasticall and Civil*, Michael Oakeshott (ed.), with an introduction by Richard S. Peters (New York: Collier Books, 1962), 103.

4. John Locke, "The Second Treatise of Government," in *Two Treatises of Government*, with an introduction and notes by Peter Laslett (New York: Mentor Books, 1965), 318.

5. Rene Descartes, "Discourse on Method," in *The Philosophical Works of Descartes*, vol. I, translated by E. S. Haldane and G. R. T. Ross (Cambridge: Cambridge University Press, 1931), 119.

6. Francis Bacon, "Novum Organum," in *Selected Writings*, Hugh G. Dick (ed.) (New York: Modern Library, 1955), 539.

7. An excellent discussion of the Progressive Era conservation movement is Samuel Hays, *Conservation and the Gospel of Efficiency* (Cambridge: Harvard University Press, 1959).

8. A study of American concerns about urban and industrial pollution in the nineteenth century is Martin Melosi (ed.), *Pollution and Reform in American Cities, 1870–1930* (Austin: University of Texas Press, 1980).

9. "The Second Treatise of Government," 329.

10. General discussions of environmental economics include A. M. Freeman, R. H. Haveman, and A. V. Kneese, *The Economics of Environmental Policy* (New York: John Wiley, 1973); William J. Baumol and Wallace E. Oates, *The Theory of Environmental Policy* (Englewood Cliffs, N.J.: Prentice-Hall, 1975); and Frederick Anderson et al., *Environmental Improvement Through Economic Incentives* (Baltimore: Johns Hopkins University Press, 1977).

11. For an economic analysis that attempts to improve ways of counting environmental costs, see J. V. Krutilla and A. C. Fisher, *The Economics of Natural Environments* (Baltimore: Johns Hopkins University Press, 1974). See also the general discussion in Joseph M. Petulla, *American Environmentalism: Values, Tactics and Priorities* (College Station: Texas A&M Press, 1980), chap. 5.

12. Larry E. Ruff, "The Economic Common Sense of Pollution," in

Edwin Mansfield (ed.), *Microeconomics: Selected Readings* (New York: W. W. Norton, 1975), 498.

13. Typical of economic thinking about "trade-offs" in environmental policy is William F. Baxter, *People or Penguins* (New York: Columbia University Press, 1974). A lively critique of cost/benefit analysis as applied to environmental decision making is presented in Mark Sagoff, "Economic Theory and Environmental Law," *Michigan Law Review* 79:1393–1419, 1981.

14. Lester C. Thurow, *The Zero-Sum Society: Distribution and the Possibilities for Economic Change* (New York: Basic Books, 1980; Penguin Books, 1981), 105, 108.

15. William J. Baumol and Wallace E. Oates, *Economics, Environmental Policy and the Quality of Life* (Englewood Cliffs, N.J.: Prentice-Hall, 1979), 282.

16. George Perkins Marsh, *Man and Nature or Physical Geography as Modified by Human Action* (New York: C. Scribner, 1864), 43.

17. A poignant treatment of this theme is Jonathan Schell, *The Fate of the Earth* (New York: Alfred A. Knopf, 1982). "Nature, once a harsh and feared master, now lies in subjection, and needs protection against man's powers. Yet because man, no matter what intellectual and technical heights he may scale, remains embedded in nature, the balance has shifted against him too, and the threat he poses to the earth is a threat to him as well." (110)

18. Some prominent works in the literature on ecological disaster are Donella H. Meadows et al., *The Limits to Growth* (New York: Universe Books, 1972); Barry Commoner, *The Closing Circle: Nature, Man and Technology* (New York: Alfred A. Knopf, 1971); Leslie L. Roos, Jr. (ed.), *The Politics of Ecosuicide* (New York: Holt, Rinehart and Winston, 1971); Joseph Eigner, "Unshielding the Sun: Environmental Effects," *Environment*, 17(3):15–18, 1975; and Robert Heilbroner, *An Inquiry into the Human Prospect* (New York: W. W. Norton, 1974).

19. *The Closing Circle*, 217–218, 229.

20. A sense of weariness about predictions of environmental hazards is expressed in Mary Douglas and Aaron Wildavsky, *Risk and Culture* (Berkeley: University of California Press, 1982).

21. *The Closing Circle*, 41; E. P. Odum, *Ecosystem Structure and Function* (Eugene: Oregon State University Press, 1972).

22. William Ophuls, *Ecology and the Politics of Scarcity* (San Francisco: W. H. Freeman, 1977), 151–152, 163.

23. A judicious introduction to this way of thinking is provided in two articles by Holmes Rolston III, "Is There an Ecological Ethic?" *Ethics* 85:93–109, 1975; and "Can and Ought We to Follow Nature?" *Environmental Ethics*, 1(2):7–30, 1979.

24. Lynn White, Jr., "The Historical Roots of Our Ecologic Crisis," *Science* 155:1203–1207, 1967.

25. Arne Naess, "The Shallow and the Deep, Long-Range Ecology Movements," *Inquiry* 16:95–100, 1973. See also Bill Devall and George Sessions, *Deep Ecology* (Layton, Utah: Gibbs M. Smith, 1984); and *Deep Ecology*,

Michael Tobias, (ed.) (San Diego: Avant, 1984).

26. René Dubos, *A God Within* (New York: Scribner's, 1972). See also Dubos' "A Theology of Earth," in Ian G. Barbour (ed.) *Western Man and Environmental Ethics* (Reading, Mass.: Addison-Wesley, 1973) and "Guest Editorial: Nature Does Not Always Know Best," in G. Tyler Miller, *Living in the Environment* (San Francisco: Wadsworth, 1975).

27. Aldo Leopold, *A Sand County Almanac* (Oxford: Oxford University Press, 1949), 204.

28. Ibid., 224–225.

29. For arguments in this vein see Peter Singer, *Animal Liberation* (New York: New York Review, 1975); and Christopher Stone, *Should Trees Have Standing? Toward Legal Rights for Natural Objects* (New York: Avon, 1975).

30. David Brower, quoted in John McPhee, *Encounters with the Archdruid* (New York: Ballantine, 1972), 65.

31. *Risk and Culture*, 158.

32. Murray Bookchin, *The Ecology of Freedom: The Emergence and Dissolution of Hierarchy* (Palo Alto: Cheshire Books, 1982), 22.

33. Anne and Paul Ehrlich, "The Population Bomb Revisited," *Life After '80: Environmental Choices We Can Live With* in Kathleen Courrier (ed.) (Andover, Mass.: Brick House, 1980), 69.

34. Georg Lukacs, *History and Class Consciousness*, translated by Rodney Livingstone (Cambridge: MIT Press, 1968), 234.

CHAPTER 8. *On Not Hitting the Tar-Baby*

1. Samuel S. Epstein, Lester O. Brown, and Carl Pope, *Hazardous Waste in America* (San Francisco: Sierra Club Books, 1982), 6.

2. Stephen Salsbury, *The State, the Investor, and the Railroad: Boston & Albany, 1825–1867* (Cambridge: Harvard University Press, 1967), 182–187.

3. Upton Sinclair, *The Jungle* (New York: Doubleday, 1906). See also Leon A. Harris, *Upton Sinclair, American Rebel* (New York: Crowell, 1975).

4. Ivan Illich, *Medical Nemesis: The Expropriation of Health* (New York: Pantheon Books, 1976), 3. "Increasing and irreparable damage," Illich writes, "accompanies present industrial expansion in all sectors. In medicine this damage appears as iatrogenesis. Iatrogenesis is clinical when pain, sickness, and death result from medical care; it is social when health policies reinforce an industrial organization that generates ill-health." (270–271)

5. See, for example, Mark Green and Robert Massie, Jr. (eds.), *The Big Business Reader* (New York: Pilgrim Press, 1980), especially Edward Greer, "And Filthy Flows the Calumet," 169–179; Michael H. Brown, "Love Canal and the Poisoning of America," 189–207; and Mark Dowie, "The Dumping of Hazardous Products on Foreign Markets," 430–443.

6. Two important writings that helped place the concept of "risk" at the center of discussions about technology, environment, and society are Chauncey Starr, "Social Benefit Versus Technological Risk," *Science* 165:1232–

1238, 1969; and William Lowrance, *Of Acceptable Risk* (Los Altos: William Kaufmann, 1976).

7. Steven D. Jellinek laments this state of affairs in "On the Inevitability of Being Wrong," *Technology Review* 82(8):8–9, 1980.

8. Dante Picciano, "Pilot Cytogenetic Study of Love Canal, New York," prepared by the Biogenics Corporation for the Environmental Protection Agency, May 14, 1980; quoted in *Hazardous Waste in America*, 113–114.

9. See Edmund A. C. Crouch and Richard Wilson, *Risk/Benefit Analysis* (Cambridge: Ballinger, 1982).

10. See Baruch Fischoff et al., *Acceptable Risk* (Cambridge: Cambridge Unviersity Press, 1981).

11. Chauncey Starr, Richard Rudman, and Chris Whipple, "Philosophical Basis for Risk Analysis," *Annual Review of Energy* 1:629–662, 1976.

12. Samuel C. Florman, "Technophobia in Modern Times," *Science '82*, 3:14, 1982.

13. Mary Douglas and Aaron Wildavsky, *Risk and Culture: The Selection of Technical and Environmental Dangers* (Berkeley: University of California Press, 1982). See my review, "Pollution as Delusion," *New York Times Book Review*, August 8, 1982, 8, 18.

14. From "Observations," an advertisement of the Mobil Oil Corporation, *Parade Magazine*, December 12, 1982, 29.

15. David Noble argues this position forcefully in "The Chemistry of Risk: Synthesizing the Corporate Ideology of the 1980s," *Seven Days* 3(7):23–26, 34, 1979. A similar view is offered by Mark Green and Norman Waitzman, "Cost, Benefit and Class," *Working Papers for a New Society*, 7(3):39–51, 1980.

16. Intentions of this sort are evident in the research on risk assessment of The Center for Technology, Environment, and Development at Clark University. See Patric Derr, Robert Goble, Roger E. Kasperson, and Robert W. Kates, "Worker/Public Protection: The Double Standard," *Environment* 23(7):6–15, 31–36, 1981; Julie Graham and Don Shakow, "Risk and Reward: Hazard Pay for Workers," *Environment* 23(8):14–20, 44–45, 1981.

17. Talbot Page discusses issues of this kind seeking a balance between "false positives and false negatives" in judgments about risks in "A Generic View of Toxic Chemicals and Similar Risks," *Ecology Law Quarterly* 7:207–244, 1978.

18. Among the helpful criticisms of cost/benefit analysis are Mark Sagoff, "Economic Theory and Environmental Law," *Michigan Law Review*, 79:1393–1419, 1981; and Steve H. Hanke, "On the Feasibility of Benefit-Cost Analysis," *Public Policy* 29:147–157, 1981.

19. Sheldon Krimsky, *Genetic Alchemy* (Cambridge: MIT Press, 1982).

20. Joel Chandler Harris, *The Complete Tales of Uncle Remus*, compiled by Richard Chase (Boston: Houghton Mifflin, 1955), 6–8.

Chapter 9. *Brandy, Cigars, and Human Values*

1. Nadezhda Mandelstam, *Hope Abandoned: A Memoir*, translated from the Russian by Max Hayward (London: Collins and Harvill Press, 1974), 10.

Chapter 10. The Whale and the Reactor

1. Henry Adams, *The Education of Henry Adams* (New York: Modern Library, 1931), 380.

2. Clark Kerr et al., *Industrialism and Industrial Man* (Cambridge: Harvard University Press, 1960). See also Clark Kerr, *The Uses of the University* (New York: Harper and Row, 1963).

3. For a chronology of events in the planning, construction, and licensing of the Diablo Canyon nuclear power plant, see Carl Neiburger "Diablo from Groundbreaking to Start-up," *San Luis Obispo County Telegram Tribune*, August 11, 1984, B7–B9.

INDEX

Index

Hard technology, concept of, 73
Harris, Joel Chandler, 154
Hart, Joseph K., 95
Harvesters, mechanical, 26–28
Haussman, Baron Eugène Georges, 23–24
Hayden, Dolores, 79
Hayes, Denis, 19–20
Hazards, environmental, 139–54
Heidegger, Martin, 4
Hennessey, James, 77–78
Hexachlorophene, 170
Hierarchy, 35, 48, 113
Highways, 22–23
Hobbes, Thomas, 52, 122, 129–30, 137, 140
Horkheimer, Max, 159
Human: condition, 13; needs, 170; values, 155–63
Humanities, 155–63
Husserl, Edmund, 22

Iatrogenic disease, 141
Icarus and Daedalus, myth of, 13
Ideals, social, 101
Ideology, 102–15, 141–42; of electronic information, 102–15
Illich, Ivan, 72, 141–42
Impacts of technological change, 10, 14, 21, 55, 100, 147–48, 176
Imperatives, technological, 29–38
In These Times, 142
Incompetence, 162
Individuality, 14–15
Industrial revolution, 43, 99
Industrialism and Industrial Man, 167
Information, electronic, 98–117; and democracy, 103–105; and ideology, 113–15; perishable commodity, 113–14; and power, 108–13, 116–17
Information society, concept of, 102–106
Information systems, 102
Inherently political technologies, 29–39, 50, 175
Innovation, 54
Instrumental understanding, 9
Instrumentality, regimes of, 54–58
Insull, Samuel, 49
Intentions, 25–26

Intermediate technology. *See* Appropriate technology
Intermediate Technology Development Group, 62, 70, 82
Invention, 22, 79
Ives, Charles, 74

Jackson, Andrew, 81
Jefferson, Thomas, 42–43, 47, 102
Jesus of Nazareth, 161
Jobs, Steven, 102
Johnson, Lyndon Baines, 167
Jones Beach, 23
Journal of the New Alchemists, 71
Jungle, The, 141
Justice, social, 99, 152

Kellogg Corporation, 93
Kerr, Clark, 167
King Lear, 122
Knowledge and power, 108–13
Koppleman, Lee, 23
Kropotkin, Peter, 19, 63, 89, 96

Labor, division of, 31–32
Land ethic, 131
Language and technology, ix, 7, 15, 49, 163
Laws, 41
Lenin, V. I., 102
Leopold, Aldo, 132, 135
Leviathan, 129
Licklider, J. C. R., 98, 105
Life, forms of. *See* Forms of life
Lillienthal, David, 20
Limits, xi, 51–58, 116–17, 155
Locke, John, 122, 124, 132, 137
Long Island, New York, highways, 22–23
Love Canal, 144
Lovins, Amory B., 53–55, 72
Los Angeles, 50, 174
Lowell, Massachusetts, 44
Lukacs, Georg, 135
Luxury, 43–44

McCormick, Cyrus II, 24–25
McLarney, Bill, 71
McNamara, Robert S., 168
Machiavelli, Niccolò, 44

196

INDEX